Applications of Global Analysis in Mathematical Physics

Jerry Marsden

University of California, Berkeley

COPYRIGHT © JERROLD MARSDEN 1974
All rights reserved.

AMS 1970 SUBJECT CLASSIFICATION:
34C40, 34D20, 35L60, 35Q10, 47B25, 47H15,
53B30, 53C30, 58D05, 58F05, 70H05, 76D10,
A1A09, 83E05.

ISBN 0-914098-11-X
Library of Congress Catalog Card Number 74-75308

PUBLISH OR PERISH, INC.
2000 CENTER ST., SUITE 1404
BERKELEY, CA. 94704 (U.S.A.)

In Japan distributed exclusively by
KINOKUNIYA BOOK-STORE CO., LTD.
TOKYO, JAPAN

APPLICATIONS OF GLOBAL ANALYSIS IN MATHEMATICAL PHYSICS

J. Marsden

Introduction
1. Infinite Dimensional Manifolds 3
2. Hamiltonian Systems 26
3. Elliptic Operators and Function Spaces 50
4. The Motion of an Incompressible Fluid 72
5. Turbulence and Chorin's formula 119
6. Symmetry Groups in Mechanics 142
7. Quantum Mechanical Systems 167
8. Completeness Theorems and Nonlinear Wave Equations 189
9. General Relativity as a Hamiltonian System 204
10. Linearization Stability of the Einstein Equations 229
Appendix: On the correspondence principle 242
Bibliography 246

APPLICATIONS OF GLOBAL ANALYSIS IN MATHEMATICAL PHYSICS

J. Marsden

Introduction.

These notes are based on a series of ten lectures given at Carleton University, Ottawa, from June 21 through July 6, 1973. The notes follow the lectures fairly closely except for a few minor amplifications.

The purpose of the lectures was to introduce some methods of global analysis which I have found useful in various problems of mathematical physics. Many of the results are based on work done with P. Chernoff, D. Ebin, A. Fischer and A. Weinstein. A more complete exposition of some of the points contained here may be found in Chernoff-Marsden [1] and Marsden-Ebin-Fischer [1] as well as in references cited later.

"Global Analysis" is a vague term. It has, by and large, two more or less distinct subdivisions. On the one hand there are those who deal with dynamical systems emphasizing topological problems such as structural stability (see Smale [2]). On the other hand there are those who deal with problems of nonlinear functional analysis and partial differential equations using techniques combining geometry and analysis. It is to the second group that we belong.

One of the first big successes of global analysis (in the second sense above) was Morse theory as developed by Palais [7] and Smale [3] and preceeded by the ideas of Leray-Schauder, Lusternik-Schnirelman and Morse. The result is a beautiful geometrization and powerful extension of the classical calculus of variations. (See Graff [1] for more up-to-date work.)

It is in a similar spirit that we proceed here. Namely we want to make use of ideas from geometry to shed light on problems in analysis which arise in mathematical physics. Actually it comes as a pleasant surprize that this point of view is useful, rather than being a mere language convenience and an outlet for generalizations. As we hope to demonstrate in the lectures, methods of global analysis can be useful in attacking specific problems.

The first three lectures contain background material. This is basic and more or less standard. Each of the next seven lectures discusses an application with only minor dependencies, except that lectures 4 and 5, and 9 and 10 form units. Lectures 4 and 5 deal with hydrodynamics and 9 and 10 with general relativity. Lecture 6 deals with miscellaneous applications, both mathematical and physical, of the concepts of symmetry groups and conserved quantities. Lecture 7 studies quantum mechanics as a hamiltonian system and discusses, e.g. the Bargmann-Wigner theorem. Finally lecture 8 studies a general method for obtaining global (in time) solutions to certain evolution equations.

It is a pleasure to thank Professors V. Dlab, D. Dawson and M. Grmela for their kind hospitality at Carleton.

1. **Infinite Dimensional Manifolds.**

Basic Calculus.

We shall let E, F, G, ... denote Banach spaces. Let $U \subset E$ be open and let $f : U \to F$ be a given mapping. We say f is <u>Fréchet differentiable at</u> $x_0 \in U$ if there is a continuous (= bounded) linear map $Df(x_0) : E \to F$ such that for all $\varepsilon > 0$ there is a $\delta > 0$ such that $\|h\| < \delta$ implies

$$\|f(x_0 + h) - f(x_0) - Df(x_0) \cdot h\| \leq \varepsilon \|h\| .$$

The map $Df(x_0)$ is necessarily unique.

Let $L(E, F)$ denote the space of all continuous linear maps from E to F together with the operator norm

$$\|T\| = \sup_{\|x\| \leq 1} \|T \cdot x\|$$

so that $L(E, F)$ is a Banach space. Let $L_s(E, F)$ denote the same space with the strong operator topology; i.e. the topology of pointwise convergence.

If f is Fréchet differentiable at each $x \in U$ and if $x \mapsto Df(x) \in L(E, F)$ (resp. $L_s(E, F)$) is continuous, we say f is of class C^1 (resp. T^1).

By induction it is not hard to formulate what it means for f to be of class C^r or T^r. For our purposes we shall be mostly dealing with C^r although T^r does arize in certain problems (see Abraham [6] and Chernoff-Marsden [2]).

The usual rules of calculus hold. Foremost amongst these is the chain rule:

$$D(f \circ g)(x) = Df(g(x)) \circ Dg(x).$$

To obtain substantial results, one often employs the following:

<u>Inverse Function Theorem.</u> <u>Let $f: U \subset E \to F$ be C^r, $r \geq 1$. Assume $Df(x_0)$ is an isomorphism for some $x_0 \in U$. Then there exists open neighborhoods U_0 of x_0 and V_0 of $f(x_0)$ such that $f: U_0 \to V_0$ is bijective and has a C^r inverse $f^{-1}: V_0 \to U_0$.</u> (We say f is a <u>local diffeomorphism</u>.)

The proof of this is essentially the same as one learns in advanced calculus where E, F are taken to be R^n. For details, see Lang [1] or Dieudonné [1].

<u>Implicit Function Theorem.</u> <u>Let $U \subset E$, $V \subset F$ be open and $f: U \times V \to G$ be C^r, $r \geq 1$. For $x_0 \in U$, $y_0 \in V$, assume $D_2 f(x_0, y_0)$ (the</u>

derivative with respect to y) _is an isomorphism of_ F _onto_ G. _Then there is a unique_ C^r _map_ $g : U_0 \times W_0 \to V$ _where_ U_0, W_0 _are sufficiently small neighborhoods of_ x_0 _and_ $f(x_0, y_0)$ _respectively, such that_

$$f(x, g(x, w)) = w$$

for all $(x, w) \in U_0 \times W_0$.

Indeed, this follows from the inverse function theorem applied to the map $\Phi : U \times V \to E \times G$

$$\Phi(x, y) = (x, f(x, y))$$

which is a local diffeomorphism.

Results like these are central to the study of submanifolds which we deal with later. They, in turn, are crucial to several of the applications.

In applications, the spaces E, F, ... are often spaces of functions and $f : E \to F$ may be some sort of non-linear differential operator. Then $Df(x_0)$ will be what is called the _linearization_ of f about x_0.

Manifolds.

Let M be a (Hausdorff) topological space. We say M is a C^∞ _manifold_ _modelled on the Banach space_ E when it has the following additional structure: there is an open covering $\{U_\alpha\}$ of M together with homeomorphisms

$$\varphi_\alpha : U_\alpha \to V_\alpha \subset E$$

where V_α is open in E such that for all α, β, the overlap map (or coordinate change)

$$\varphi_\alpha \circ \varphi_\beta^{-1}$$

(defined on $\varphi_\beta(U_\alpha \cap U_\beta)$) is a C^∞ map.

By a <u>chart</u> (or <u>coordinate patch</u>) we mean a homoemorphism $\varphi : U \subset M \to V \subset E$ of open sets such that for all α, the map

$$\varphi \circ \varphi_\alpha^{-1}$$

(defined on $\varphi_\alpha(U_\alpha \cap U)$) is C^∞.

The collection of all charts yields what is called a <u>maximal atlas</u>.

Let M and N be manifolds and $f : M \to N$ a continuous map. We say f is of <u>class</u> C^r if for every chart $\varphi : U \subset M \to V \subset E$ of M and $\psi : U_1 \subset N \to V_1 \subset F$ of N, the map

$$\psi \circ f \circ \varphi^{-1}$$

of the open set $\varphi(f^{-1}(U_1) \cap U)$ to F is C^r. By the chain rule one sees that this holds for all charts if it holds for some covering of N and M by charts.

Submanifolds.

Let M be a manifold and let $S \subset M$. In applications S is often defined by some restrictive condition; e.g. by constraints of the form $f(x) = c$. It is important to know whether or not S is smooth, e.g. has no sharp corners. Below we give a useful condition for this, but first let us formulate the definition.

We say S is a submanifold of M (where M is modelled on E) if we can write $E = F \oplus G$ (topological sum) and for every $x \in S$, there is a chart $\varphi : U \subset M \to V \subset E$ of M where $x \in U$ such that

$$\varphi(U \cap S) = V \cap (F \times \{w\})$$

where $w \in G$.

In other words, the chart φ "flattens out" S making it lie in the subspace F.

One sees that the above charts define a manifold structure for S; S will be modelled on F. The conditions ensure that the manifold structure on S is compatible with that on M.

Vector Bundles.

By a vector bundle we mean a manifold E together with a

submanifold $M \subset E$ and a projection $\pi : E \to M$ (i.e. $\pi \circ \pi = \pi$) such that for each $x \in M$, the fibre $E_x = \pi^{-1}(x)$ is a linear space with x as the zero element; there should also be a covering by charts (called vector bundle charts) of the form

$$\varphi : \pi^{-1}(U) \subset E \to V \times F$$

where U is open in M, $V \subset E$, the model space for M and F is some fixed Banach space such that the overlap maps are linear isomorphisms when restricted to each fiber.

Intuitively, one thinks of a vector bundle over M as a collection of linear spaces E_x, one attatched to each $x \in M$. For $v \in E_x$, $\pi(v) = x$ is the base point to which v is attatched. As we shall see, the quantities v can be vectors, tensors, differential forms, spinors, etc.

The Tangent Bundle.

The most basic vector bundle attatched to a manifold M is its tangent bundle TM. It was an important observation in the historical development of manifold theory, that the tangent space to a manifold can be defined completely intrinsically. For example there is no need to have a space in which the manifold is embedded; one might think such an embedding is necessary by thinking of surfaces in R^3.

There are two useful and equivalent ways to define T_xM, the fibre of TM above $x \in M$.

First, we can use curves. Indeed, intuitively a tangent vector $v \in T_x M$ ought to be $c'(0)$ for some curve $c(t)$ in M with $c(0) = x$. So consider all curves $c : R \to M$ with $c(0) = x$ and say $c_1 \sim c_2$ if $c_1'(0) = c_2'(0)$ in some (and hence every) chart about x, where $c' = dc/dt$ in that chart. Then $T_x M$ is defined to be the set of equivalence classes, and TM is the disjoint union of the $T_x M$'s.

The above definition is useful because it is closely connected with our intuition. There is a second definition which brings out the vector bundle structure of TM more clearly. This goes as follows.

Fix $x \in M$ again and look at charts φ defined on neighborhoods of x. Consider pairs (φ, e) where $e \in E$, the model space of M. Say

$$(\varphi_1, e_1) \sim (\varphi_2, e_2)$$

if

$$D(\varphi_2 \circ \varphi_1^{-1})(\varphi_1(x)) \cdot e_1 = e_2$$

Then $T_x M$ is the set of equivalence classes of such pairs. Clearly $T_x M$ is a linear space. Moreover a chart φ induces naturally a vector bundle chart on TM by using the definition, and these charts make M manifestly a submanifold.

We leave it to the reader to check the equivalence of the

two definitions. We use $\pi : TM \to M$ for the projection.

One often uses definitions involving derivations for $T_x M$ in the case of finite dimensional manifolds. For infinite dimensional manifolds this is possible but is rather cumbersome (see Schwartz [1], p. 105 for a discussion).

Let $f : M \to N$ be a C^r map, $r \geq 1$. Then there is a bundle map $Tf : TM \to TN$ naturally induced; i.e. Tf maps fibres to fibres and the following diagram commutes:

$$\begin{array}{ccc} TM & \xrightarrow{Tf} & TN \\ \pi \downarrow & & \downarrow \pi \\ M & \xrightarrow{f} & N \end{array}.$$

Commutativity of this diagram means nothing more than for $x \in M$ 6, and $v \in T_x M$, $Tf(v) \in T_{f(x)} N$.

Using the first definition of TM, we define $Tf(v) = \frac{d}{dt} f(c(t))\big|_{t=0}$ where $v = c'(0)$. (Remember $c'(0)$ stands for the equivalence class of curves and in a chart for TM, it really is the derivative.)

Actually this definition is very useful for doing computations, as we shall see later.

Using the second definition, if the local representative for f is $f_{\varphi, \psi} : V \subset E \to V_1 \subset F$ relative to charts φ on M and

ψ on N, then the local representative for Tf is, in the corresponding charts for TM and TN,

$$(Tf)_{\varphi,\psi} : V \times E \to V_1 \times F$$
$$(x, e) \mapsto (f_{\varphi\psi}(x), Df_{\varphi,\psi}(x) \cdot e)$$

One checks that this is consistent with the equivalence relation and so yields a well defined map Tf.

In the language of tangents, the chain rule can be neatly expressed by saying that

$$T(f \circ g) = Tf \circ Tg .$$

Submersions.

Let E be a topological vector space, and $F \subset E$ a closed subspace. We say F <u>splits</u> if there is another closed subspace G such that

$$E = F \oplus G \quad \text{(topological sum)} .$$

For example if E is a Hilbert space this is always the case, for we can choose $G = F^\perp$. However in a general Banach space a closed subspace need not have a closed complement.

We say topological vector space rather than Banach space here because we want to use the case where E is T_xM. The latter does not carry canonically the structure of a Banach space, but it does have the structure of a topological vector space (if a norm is assigned to each tangent space $T_x M$, one speaks of a <u>Finsler structure</u>).

Now let M and N be Banach manifolds and $f : M \to N$ a C^∞ map. We want to know when $S = f^{-1}(w)$ is a submanifold of M, where $w \in N$ is fixed. We say f is a <u>submersion</u> on S if for all $x \in S$, $T_x f : T_x M \to T_{f(x)} N$ is surjective and kernel $T_x f$ splits.

<u>Theorem</u>. <u>Let f be a submersion on S as just described. Then S is a smooth submanifold of M</u>.

<u>Proof</u>. Work in a chart $U \subset E$ for M. Write $E = E_0 \oplus E_1$ where $E_0 = \ker Df(x)$, for x fixed. Consider the map Φ defined near x to $E_0 \times F$ by

$$\Phi(x_0, x_1) = (x_0, f(x_0, x_1)).$$

Since E_0 is kernel $Df(x)$ we see that $D_2 f$ is an isomorphism from E_1 to F and so $D\Phi$ at x is an isomorphism. The map Φ is therefore a local diffeomorphism by the inverse function theorem. Clearly Φ yields a chart showing S is a submanifold. □

Since S is modelled on E_0, this argument also shows:

<u>Corollary</u>. $T_x S = \text{kernel } T_x f$.

To make effective use of this result one must be judicious

in the choice of N. The space N must be large enough so f maps into N, but only just large enough to ensure that $T_x f$ will be surjective.

There is a similar result for <u>immersions</u>. Here $f : M \to N$ should be injective and have injective tangent at each $x \in M$ and the image should split. Also, the map f should be closed. Then f(M) will be a submanifold of N.

The reader can work this case out for himself. We have stressed the submersion case because it is more useful for the sort of applications that we have in mind.

Differential Forms.

Given a linear topological space E, we let E^* denote the dual space; i.e. the space of all continuous linear maps $\ell : E \to R$.

Let M be a manifold and TM its tangent bundle. We can form a new bundle T^*M over M whose fibre over $x \in M$ is the dual space T_x^*M. It is not hard to see that this is a vector bundle. It is called the <u>cotangent bundle</u>.

In general if $\pi : E \to M$ is a vector bundle, a <u>section</u> s of E is a map

$$s : M \to E$$

such that

$$\pi \circ s = \text{identity} .$$

In other words, $s(x) \in E_x$ for each $x \in M$.

A section X of the tangent bundle is called a <u>vector field</u> while a section α of the cotangent bundle is called a <u>one-form</u> or a <u>covector field</u>.

Let $f : M \to R$. Then since $TR = R \times R$, $T_x f : T_x M \to R$, or in other words $T_x f \in T_x^* M$. Thus the tangent of f naturally induces a one form on M. So regarded, it is denoted df and is called the <u>differential</u> of f.

We can generalize T^*M as follows. Let $\wedge^k M$ be the vector bundle over M whose fiber at $x \in M$ is the k-multilinear alternating continuous maps $T_x M \times \ldots \times T_x M \to R$. A section of the bundle $\wedge^k M$ is called a k-<u>form</u>. We regard real-valued functions as 0-forms.

Let α be a k-form and β an ℓ-form. Then the <u>wedge product</u> $\alpha \wedge \beta$ is defined as

$$(\alpha \wedge \beta)_x(v_1,\ldots,v_{k+\ell}) = \sum_\sigma (\text{sgn } \sigma) \alpha_x(v_{\sigma(1)},\ldots,v_{\sigma(k)}) \beta_x(v_{\sigma(k+1)},\ldots,v_{\sigma(k+\ell)})$$

where the sum is over all permutations σ such that $\sigma(1) < \ldots < \sigma(k)$ and $\sigma(k+1) < \ldots < \sigma(k+\ell)$ and sgn $\sigma = \pm 1$ is the sign of σ.

Note: We use the conventions of Bourbaki [1]. Compare with Abraham [2].

In the case of R^3 we can identify one forms and two forms with vectors. When we do so, $\alpha \wedge \beta$ is seen to be just the cross product.

If $f : M \to N$ is a C^r mapping and α is a C^r k-form on N we get a C^{r-1} k-form $f^*\alpha$ on M defined by

$$(f^*\alpha)_x(v_1, \ldots, v_k) = \alpha_{f(x)}(Tf \cdot v_1, \ldots, Tf \cdot v_k) .$$

We call $f^*\alpha$ the <u>pull back</u>* of α by f.

If $X_1, \ldots, X_k$ are vector fields on M and α is a k-form we get a real valued function $\alpha(X_1, \ldots, X_k)$ defined by

$$\alpha(X_1, \ldots, X_k)(x) = \alpha_x(X_1(x), \ldots, X_k(x)) .$$

Notice that the differential mapped a 0-form to a 1-form. This can be generalized as follows. If α is a k-form, define the $k + 1$ form $d\alpha$ by

$$d\alpha_x(v_0, \ldots, v_k) = \sum_{i=0}^{k} (-1)^i (D\alpha_x \cdot v_i)(v_0, \ldots, \hat{v}_i, \ldots, v_k)$$

where $\hat{v}_i$ denotes that v_i is missing and $D\alpha_x$ is the derivative of α in charts; note $\alpha : U \subset E \to \wedge^k(E)$ so $D\alpha_x : E \to \wedge^k(E)$. One can check that d is chart independent. The operator d plays a fundamental role in calculus on manifolds. It is called the <u>exterior derivative</u>.

It is not hard, but a little tedious, to verify:

---

*Note that the positioning of the stars agrees with Bourbaki[1], Lang [1] but is the opposite of Abraham [2].

(i) d is real linear

(ii) d∘d = 0

and (iii) $d(\alpha \wedge \beta) = d\alpha \wedge \beta + (-1)^k \alpha \wedge d\beta$.

Condition (ii) is a generalization of the familiar identity $\nabla \times (\nabla f) = 0$ from vector analysis.

If α is a k-form and X a vector field, define the <u>interior product</u> $i_X \alpha = X \lrcorner \alpha$ by

$$(i_X \alpha)_x(v_2, \ldots, v_k) = \alpha_x(X(x), v_2, \ldots, v_k)$$

so $i_X \alpha$ is a k-1 form.

Define the <u>Lie derivative</u> $L_X \alpha$ by

$$L_X \alpha = d i_X \alpha + i_X d\alpha$$

so $L_X \alpha$ is a k-form if α is. This object is an extremely useful tool in Hamiltonian mechanics as we shall soon see.

Following are some of the basic operations on vector fields. If X is a vector field on M and $f : M \to \mathbb{R}$, we set

$$X(f) : M \to \mathbb{R} , \quad x \mapsto df_x \cdot X(x) .$$

By convention $i_X f = 0$, so we see that $X(f) = L_X f$. Notice that $X(f)$ is nothing but the derivative of f in the direction of X.

Clearly $X(f)$ is a derivation in f : $X(fg) = fX(g) + gX(f)$.
As we have mentioned, this property is sometimes used (in finite dimensions) to characterize vector fields.

Let X and Y be vector fields. Then their bracket $[X, Y]$ is a vector field on M such that in local coordinates

$$[X, Y] = DY \cdot X - DX \cdot Y .$$

As a derivation, we have

$$[X, Y](f) = X(Y(f)) - Y(X(f)) .$$

Let f be a C^r diffeomorphism of M onto N, $r \geq 1$. That is, f is C^r, a bijection with C^r inverse. Given a vector field X on M, set

$$f_* X = Tf \circ X \circ f^{-1}$$

a vector field on N. Similarly if Y is a vector field on N, set

$$f^* Y = Tf^{-1} \circ Y \circ f .$$

These are characterised by

$$(f_* X)(h) = (X(h \circ f)) \circ f^{-1}$$

which follows from the chain rule.

In the following table we summarize some of the useful identities connecting the various operations which we have introduced. The proofs are straightforward algebraic manipulations. These identities are quite convenient in various applications as we shall see in the next lecture.

1. Vector fields on M with the bracket $[X, Y]$ form a Lie algebra; i.e., $[X, Y]$ is real bilinear, skew symmetric and Jacobi's identity holds: $[[X, Y], Z] + [[Z, X], Y] + [[Y, Z], X] = 0$.

2. For a diffeomorphism f, $f_*[X, Y] = [f_*X, f_*Y]$ and $(f \circ g)_*X = f_* g_* X$.

3. The forms on a manifold are a real associative algebra with $\wedge$ as multiplication. Furthermore, $\alpha \wedge \beta = (-1)^{k\ell} \beta \wedge \alpha$ for k and ℓ forms α and β respectively.

4. If f is a map, $f^*(\alpha \wedge \beta) = f^*\alpha \wedge f^*\beta$, $(f \circ g)^*\alpha = g^* f^* \alpha$.

5. d is a real linear map on forms and: $dd\alpha = 0$, $d(\alpha \wedge \beta) = d\alpha \wedge \beta + (-1)^k \alpha \wedge d\beta$ for α a k-form.

6. For α a k-form and $X_0, \ldots, X_k$ vector fields:

$$d\alpha(X_0, \ldots, X_k) = \sum_{i=0}^{k} (-1)^i X_i(\alpha(X_0, \ldots, \hat{X}_i, \ldots, X_k)) +$$

$$+ \sum_{i<j} (-1)^{i+j} \alpha([X_i, X_j], X_0, \ldots, \hat{X}_i, \ldots, \hat{X}_j, \ldots, X_k).$$

7. For a map f, $f^*d\alpha = df^*\alpha$.

8. (Poincaré lemma). If $d\alpha = 0$ then α is locally exact; i.e. there is a neighborhood U about each point on which $\alpha = d\beta$.

9. $i_X\alpha$ is real bilinear in X, α and for $h: M \to \mathbb{R}$, $i_{hX}\alpha = hi_X\alpha = i_X h\alpha$. Also $i_X i_X \alpha = 0$, $i_X(\alpha \wedge \beta) = i_X\alpha \wedge \beta + (-1)^k \alpha \wedge i_X\beta$.

10. For a diffeomorphism f, $f^* i_X \alpha = i_{f^*X} f^*\alpha$.

11. $L_X\alpha = di_X\alpha + i_X d\alpha$.

12. $L_X\alpha$ is real bilinear in X, α and $L_X(\alpha \wedge \beta) = L_X\alpha \wedge \beta + \alpha \wedge L_X\beta$.

13. For a diffeomorphism f, $f^* L_X \alpha = L_{f^*X} f^*\alpha$.

14. $(L_X\alpha)(X_1,\ldots,X_k) = X(\alpha(X_1,\ldots,X_k)) - \sum_{i=1}^{k} \alpha(X_1,\ldots,[X,X_i],\ldots,X_k)$.

15. Locally, $(L_X\alpha)_x \cdot (v_1,\ldots,v_k) = D\alpha_x \cdot X(x) \cdot (v_1,\ldots,v_k)$

$$+ \sum_{i=1}^{k} \alpha_x \cdot (v_1,\ldots,DX_x \cdot v_i,\ldots,v_k).$$

16. The following identities hold:

$$L_{fX}\alpha = fL_X\alpha + df \wedge i_X\alpha$$

$$L_{[X,Y]}\alpha = L_X L_Y\alpha - L_Y L_X\alpha$$

$$i_{[X,Y]}\alpha = L_X i_Y\alpha - i_Y L_X\alpha$$

$$L_X d\alpha = dL_X\alpha$$

$$L_X i_X\alpha = i_X L_X\alpha.$$

TABLE 1

Flows of Vector Fields.

By a *flow* (or a one parameter group of diffeomorphisms) we mean a collection of (smooth) maps $F_t : M \to M$, $t \in \mathbb{R}$ such that

$$\begin{cases} F_{t+s} = F_t \circ F_s \\ F_0 = \text{identity} \end{cases}.$$

The term *dynamical system* is also used. For each $x \in M$, $t \mapsto F_t(x)$ is the *trajectory* of x. The condition $F_{t+s} = F_t \circ F_s$ expresses nothing more than "causality".

If X is a vector field, we say it *has flow* F_t if

$$\frac{d}{dt} F_t(x) = X(F_t(x)).$$

In other words, $F_t(x)$ solves the system of differential equations determined by X. In finite dimensions these are ordinary differential equations. In infinite dimensions certain types of partial differential equations can be handled; we shall discuss this point below.

If the individual solution curves (or integral curves) of X are unique; i.e. if

$$\frac{d}{dt} c(t) = X(c(t))$$

$$c(0) = x_0$$

has a unique solution, then one can prove the above flow property

$F_{t+s} = F_t \circ F_s$ rather easily.

An important point is that, in general, the flow of a vector field need not be defined for all $t \in R$ for each $x \in M$. For example a trajectory can leave the manifold in a finite time. To ensure this doesn't happen, one requires certain estimates to establish that a trajectory remains in a bounded region for bounded t-intervals. If $F_t(x)$ is defined for all $t \in R$ we say X has a complete flow.

Local Existence and Uniqueness Theorem. If X is a C^r vector field, $r \geq 1$, then X has a locally defined, unique C^r flow F_t.

This result is proved by the Picard iteration method, as one learns in elementary courses on differential equations. Thus it is similar to the proof of the inverse function theorem. Actually it can be deduced from the latter directly; see Robbin [1]. (In fact the other cornerstone of differential analysis, the Frobenius theorem can also be so deduced; see Penot [4]).

One usually proceeds by using the above theorem to deduce the existence of the local flow and then use special properties of X to prove completeness. (For example we shall do this in certain cases for Hamiltonian vector fields.)

The Heat Equation.

In these lectures we are concerned to a great extent with partial differential equations. For these, the above theorem is rather limited. To see why, consider the heat equation:

$$\frac{\partial u}{\partial t} = \Delta u$$

where u is a function of $x \in R$ and $t \in R$ and u is given at time $t = 0$. Here $\Delta = \partial^2/\partial x^2$ is the Laplacian.

Of course this equation can be solved explicitly:

$$u(x, t) = \frac{1}{\sqrt{4t\pi}} \int_{-\infty}^{\infty} e^{-|x-y|^2/4t} u_0(y) dy , \quad t > 0 .$$

The solution is actually good only for $t \geq 0$. Nevertheless it yields a well defined <u>continuous</u> semi flow F_t on $L_2(R)$. In general it won't be t-differentiable for all $u_0 \in L_2(R)$.

Indeed Δ is not a bounded operator on $L_2(R)$ so we cannot use the existence theorem. Rather Δ is defined only on a domain $D \subset L_2(R)$, a dense linear subspace consisting of those $f \in L_2$ whose derivatives (in the distribution sense) of order ≤ 2 also lie in L_2.

One cannot remedy the situation by passing to the C^∞ functions. Indeed this is not a Banach but a Fréchet space and it is not hard to show that for these spaces the local existence theorem is false.

However there is a general theorem which can cover the situation, called the Hille Yosida theorem. We don't want to go into this now, so we just state two useful special cases: (See Yosida [1] for details. The "Schrodinger case" - Stone's theorem - will be considered later).

Parabolic Case. Let H be a Hilbert space and $A : D \subset H \to H$ a (linear) self adjoint operator, $A \leq 0$. Then the equation $\frac{du}{dt} = Au$ defines a unique linear semi-flow F_t, $t \geq 0$ on H. The equation $\frac{du}{dt} = Au$ is satisfied for $u_0 \in D$ and $u_t = F_t(u_0)$.

For example, this covers the case of the heat equation. For the wave equation, $\frac{\partial^2 u}{\partial t^2} = \Delta u$, we use:

Hyperbolic Case. Let H, A be as above. Then the equation

$$\begin{cases} \frac{du}{dt} = v \\ \frac{dv}{dt} = Au \end{cases}$$

defines a unique flow F_t, $t \in \mathbb{R}$ on $D \times H$.

Densely Defined Vector Fields.

In view of the above examples, it is useful to extend our notions about vector fields so as to include more interesting examples.

By a manifold domain $D \subset M$, we mean a dense subset D in a manifold M such that D is also a manifold and the inclusion $i : D \to M$ is smooth and Ti has dense range.

By a densely defined vector field, we mean a map $X : D \to TM$ such that for $x \in D$, $X(x) \in T_xM$. A flow (or semi-flow) for X will be a collection of maps $F_t : D \to D$, $t \in R$ (or $t \geq 0$) [perhaps locally defined] such that

$$F_{t+s} = F_t \circ F_s$$

$$F_0 = \text{identity}$$

and for $x \in D$,

$$\frac{d}{dt} F_t(x) = X(F_t(x))$$

where $\frac{d}{dt}$ is taken regarding $F_t(x)$ as a curve in M.

Such a generalization allows the more interesting examples like the heat and wave equation and non-linear generalizations of them to be included.

What about a local existence theorem? Since we already have a good theorem (the Hille-Yosida theorem) available for the linear case, it is natural to linearize. There is a theorem, the Nash-Moser theorem which is suitable for these purposes. The exact hypotheses are too complicated to give in full here, but basically the spaces must be "decent" (technically, they must admit "smoothing operators") and if X is the vector field: $X : D \subset E \to E$, the linear operators $DX(x) : E \to E$ must have flows (or semi-flows) which vary smoothly with x. Then X will have a local flow.

For further information, see J. T. Schwartz [1], J. Marsden [1] and M. L. Gromov [1]. There are also a number of special techniques available, some of which are discussed later.

Flows and Lie Derivatives.

There is a very fundamental link between the flow of a vector field and the Lie derivative.

Theorem. *Let* X *be a* C^r *vector field on* M, $r \geq 1$ *and* α *a* k *form on* M. *Let* F_t *be the flow of* X. *Then*

$$\frac{d}{dt} F_t^* \alpha = F_t^* (L_X \alpha) .$$

Actually the proof is very simple, once we have the identities in table 1. Indeed if we differentiate in a chart

$$\frac{d}{dt} (F_t^* \alpha)_x (v_1, \ldots, v_k)$$
$$= \frac{d}{dt} \alpha_{F_t(x)} (D \cdot F_t \cdot v_1, \ldots, DF_t \cdot v_k)$$

we get exactly the expression for $L_X \alpha$ in formula 15 of table 1.

For example if $L_X \alpha = 0$ then $F_t^* \alpha = \alpha$; i.e. α is preserved by the flow.

The above theorem extends also to densely defined vector fields. We just need that each $F_t : D \to D$ be C^1 and that α be smooth on M (rather than on D).

One of the points we wish to make here is that these geometrical ideas, culminating for example in the above theorem, can be applied to partial differential as well as to ordinary differential equations.

2. Hamiltonian Systems.

This lecture contains some of the basic facts about Hamiltonian systems. Some additional material will be brought in later as it is needed.

Motivation.

To motivate the development, let us briefly consider Hamilton's equations. The starting point is Newton's second law which states that a particle of mass $m > 0$ moving in a potential $V(x)$, $x \in R^3$ moves along a curve $x(t)$ such that $m\ddot{x} = -\text{grad } V(x)$. If we introduce the momentum $p = m\dot{x}$ and the energy $H(x, p) = \frac{1}{2m}\|p\|^2 + V(x)$ then Newton's law becomes __Hamilton's Equations__

$$\begin{cases} \dot{x}^i = \partial H/\partial p_i \\ \dot{p}_i = -\partial H/\partial q^i \end{cases} \quad i = 1, 2, 3.$$

One now is interested in studying this system of first order equations for given H. To do this, we introduce the matrix $J = \begin{pmatrix} 0 & I \\ -I & 0 \end{pmatrix}$ where I is the $n \times n$ identity and note that the equations become $\dot{\xi} = J \text{ grad } H(\xi)$ where $\xi = (x, p)$. (In complex notation, setting $z = x + ip$, they may be written as $\dot{z} = 2i\partial H/\partial \bar{z}$).

Suppose we make a change of coordinates $w = f(\xi)$ where $f : R^{2n} \to R^{2n}$ is smooth. If $\xi(t)$ satisfies Hamilton's equations, the equations satisfied by $w(t)$ are $\dot{w} = A\dot{\xi} = AJ \text{ grad}_\xi H(\xi) = AJA^* \text{grad}_w H(\xi(w))$ where $A = (\partial w^i/\partial \xi^j)$ is the Jacobian of f. The equations for w will be Hamiltonian with energy $K(w) = H(\xi(w))$ if $AJA^* = J$. A transformation satisfying this condition is called __canonical__ or __symplectic__.

The space $R^3 \times R^3$ of the ξ's is called the __phase space__. For a system of N particles we would use $R^{3N} \times R^{3N}$.

We wish to point out that for many fundamental physical systems, the phase space is a manifold rather than Euclidean space. These arize when constraints are present. For example the phase space for the motion of the rigid body is the tangent bundle of the group $SO(3)$ of 3×3 orthogonal matrices with determinant $+1$.

To generalize the notion of a Hamiltonian system, we first need to geometrize the symplectic matrix J above. In infinite dimensions there is a technical point however which is important. We give a discussion of this in the following.

Strong and Weak Nondegenerate Bilinear Forms.

Let E be a Banach space and $B : E \times E \to R$ a continuous bilinear mapping. Then B induces a continuous map $B^\flat : E \to E^*$, $e \mapsto B^\flat(e)$ defined through $B^\flat(e)f = B(e, f)$. We call B __weakly nondegenerate__ if $B^\flat$ is injective; i.e. $B(e, f) = 0$ for all $f \in E$ implies $e = 0$. We call B __nondegenerate__ or __strongly nondegenerate__ if $B^\flat$ is an isomorphism. By the open mapping theorem it follows that B is nondegenerate iff B is weakly nondegenerate and $B^\flat$ is onto.

If E is finite dimensional there is no difference between strong and weak nondegeneracy. However in infinite dimensions the distinction is important to bear in mind.

Let M be a Banach manifold. By a <u>weak Riemannian structure</u> we mean a smooth assignment $x \mapsto <,>_x$ of a weakly nondegenerate inner product (not necessarily complete) to each tangent space $T_x M$. Here smooth means that in local charts $x \in U \subset E \mapsto <,>_x \in L_2(E \times E, R)$ is smooth where $L_2(E \times E, R)$ denotes the Banach space of bilinear maps of $E \times E$ to R. Equivalently $<,>_x$ is a smooth section of the vector bundle whose fiber at $x \in M$ is $L_2(T_x M \times T_x M, R)$.

By a <u>Riemannian manifold</u> we mean a weak Riemannian manifold in which $<,>_x$ is nondegenerate. Equivalently, the topology of $<,>_x$ is complete on $T_x M$, so that the model space E must be isomorphic to a Hilbert space.

For example the L_2 inner product $<f, g> = \int_0^1 f(x) g(x) dx$ on $E = C([0,1], R)$ is a weak Riemannian metric on E but not a Riemannian metric.

Symplectic Forms.

Let P be a manifold modelled on a Banach space E. By a <u>symplectic form</u> we mean a two form ω on P such that

(a) ω is closed; $d\omega = 0$

(b) for each $x \in P$, $\omega_x : T_x P \times T_x P \to R$ is nondegenerate.

If ω_x in (b) is weakly nondegenerate, we speak of a <u>weak symplectic form</u>.

The need for weak symplectic forms will be clear from examples given below. For the moment the reader may wish to assume P is finite dimensional in which case the distinction vanishes.

If (b) is dropped we refer to ω as a <u>presymplectic form</u>. This case will be referred to later. The first result is referred to as <u>Darboux's theorem</u>. Our proof follows Weinstein [1]. The method is also useful in Morse theory; see Palais [5].

<u>Theorem</u>. <u>Let</u> ω <u>be a symplectic form on the Banach manifold</u> P. <u>For each</u> $x \in P$ <u>there is a local coordinate chart about</u> x <u>in which</u> ω <u>is constant</u>.

<u>Proof</u>. We can assume $P = E$ and $x = 0 \in E$. Let ω_1 be the constant form equalling $\omega_0 = \omega(0)$. Let $\tilde{\omega} = \omega_1 - \omega$ and $\omega_t = \omega + t\tilde{\omega}$, $0 \leq t \leq 1$. For each t, $\omega_t(0) = \omega(0)$ is nondegenerate. Hence by openness of the set of linear isomorphisms of E to E^*, there is a neighborhood of 0 on which ω_t is nondegenerate for all $0 \leq t \leq 1$. We can assume that this neighborhood is a ball. Thus by the Poincaré lemma (appendix 1) $\tilde{\omega} = d\alpha$ for some one form α. We can suppose $\alpha(0) = 0$.

Define a vector field X_t by $i_{X_t} \omega_t = -\alpha$ which is possible since ω_t is nondegenerate. Moreover, X_t will be smooth. Since

$X_t(0) = 0$ we can, from the local existence theory restrict to a sufficiently small ball on which the integral curves will be defined for a time at least one.

Now let F_t be the flow of X_t. The connection between Lie derivatives and flows still holds for time dependent vector fields, so we have

$$\frac{d}{dt}(F_t^* \omega_t) = F_t^*(L_{X_t} \omega_t) + F_t^* \frac{d}{dt}\omega_t$$

$$= F_t^* di_{X_t} \omega_t + F_t^* \tilde{\omega}$$

$$= F_t^*(d(-\alpha) + \tilde{\omega}) = 0 .$$

Therefore, $F_1^* \omega_1 = F_0^* \omega_0 = \omega$, so F_1 provides the chart transforming ω to the constant form ω_1. □

Of course such a result cannot be true for riemannian structures (otherwise they would be flat). Darboux's theorem is not true for weak symplectic forms. See Marsden [4]. Recently A. Tromba has found some useful sufficient conditions to cover the weak case.

<u>Corollary</u>. <u>If</u> P <u>is finite dimensional and</u> ω <u>is a symplectic form</u> <u>then</u>

    (a) P <u>is even dimensional, say</u> $\dim P = m = 2n$

    (b) <u>locally about each point there are coordinates</u> $x^1, \ldots, x^n, y^1, \ldots, y^n$ <u>such that</u>

$$\omega = \sum_{1}^{n} dx^i \wedge dy^i.$$

Such coordinates are called _canonical_.

Proof. By elementary linear algebra, any skew symmetric bilinear form which is nondegenerate has the canonical form $\begin{pmatrix} 0 & I \\ -I & 0 \end{pmatrix}$ where I is the n × n identity. This is the matrix version of (b) pointwise on M. The result now follows from Darboux's theorem. □

The corollary actually has a generalization to infinite dimensions. Clearly it is just a result on the canonical form of a skew symmetric bilinear mapping. First some notation. Let E be a real vector space. By a _complex structure_ on E we mean a linear map $J : E \to E$ such that $J^2 = -I$. By setting $ie = J(e)$ one then gives E the structure of a complex vector space. We now show that a symplectic form is the imaginary part of an inner product. (cf. Cook [1]).

Proposition. _Let_ H _be a real Hilbert space and_ B _a skew symmetric weakly nondegenerate bilinear form on_ H. _Then there exists a complex structure_ J _on_ H _and a real inner product_ s _such that_

$$s(x, y) = B(Jx, y).$$

Setting

$$h(x, y) = s(x, y) + iB(x, y),$$

h _is a hermetian inner product._ Finally, h _or_ s _is complete on_

H **iff** B **is nondegenerate.**

<u>Proof.</u> Let $<,>$ be the given complete inner product on H. By the Riesz theorem, $B(x, y) = <Ax, y>$ for a bounded linear operator $A : H \to H$. Since B is skew, we find $A^* = -A$.

Since B is weakly nondegenerate, A is injective. Now $-A^2 \geq 0$, and from $A = -A^*$ we see that A^2 is injective. Let P be a symmetric non-negative square root of $-A^2$. Hence P is injective. Since $P = P^*$, P has dense range. Thus P^{-1} is a well defined unbounded operator. Set $J = AP^{-1}$, so that $A = JP$. From $A = -A^*$ and $P^2 = -A^2$, we find that J is orthogonal and $J^2 = -1$. Thus we may assume J is a bounded operator. Moreover J is symplectic in the sense that $B(Jx, Jy) = B(x, y)$. Define $s(x, y) = B(Jx, y) = <Px, y>$ since $A = JP = PJ$. Thus s is an inner product on H. Finally, it is a straightforward check to see that h is a hermetian inner product. For example; $h(ix, y) = s(Jx, y) + iB(Jx, y) = -B(x, y) + is(x, y) = ih(x, y)$. The proposition follows. □

<u>Canonical Symplectic Forms.</u>

We recall that a Banach space E is reflexive iff the canonical injection $E \to E^{**}$ is onto. For instance any finite dimensional or Hilbert space is reflexive. The L_p spaces, $1 < p < \infty$ are reflexive, but $C([0, 1], R)$ with the sup norm is not.

Let M be a manifold modelled on a Banach space E. Let

T^*M be its cotangent bundle, and $\tau^* : T^*M \to M$ the projection. Define the <u>canonical one form</u> θ on T^*M by

$$\theta(\alpha_m) W = -\alpha_m T\tau^*(W)$$

where $\alpha_m \in T_m^*M$ and $W \in T_{\alpha m}(T^*M)$. In a chart $U \subset E$, this formula is the same as saying

$$\theta(x, \alpha) \cdot (e, \beta) = -\alpha(e)$$

where $(x, \alpha) \in U \times E^*$, $(e, \beta) \in E \times E^*$. If M is finite dimensional, this says

$$\theta = -\sum p_i dq^i$$

where $q^1, \ldots, q^n, p_1, \ldots, p_n$ are coordinates for T^*M.

The <u>canonical two form</u> is defined by $\omega = d\theta$. Locally, using the formula for d from table one, p. 19,

$$\omega(x, \alpha) \cdot ((e_1, \alpha_1), (e_2, \alpha_2)) = \{\alpha_2(e_1) - \alpha_1(e_2)\}$$

or, in the finite dimensional case,

$$\omega = \sum dq^i \wedge dp_i .$$

<u>Proposition</u> (a) <u>The form</u> ω <u>is a weak symplectic form on</u> $P = T^*M$

(b) ω <u>is symplectic iff</u> E <u>is reflexive</u>.

<u>Proof</u>. (a) Suppose $\omega(x, \alpha)((e_1, \alpha_1),(e_2, \alpha_2)) = 0$ for all (e_2, α_2).

Setting $e_2 = 0$ we get $\alpha_2(e_1) = 0$ for all $\alpha_2 \in E^*$. By the Hahn-Banach theorem, this implies $e_1 = 0$. Setting $\alpha_2 = 0$ we get $\alpha_1(e_2) = 0$ for all $e_2 \in E$, so $\alpha_1 = 0$.

(b) Suppose E is reflexive. We must show that the map $\omega^b : E \times E^* \to (E \times E^*)^* = E^* \times E^{**}$, $\omega^b(e_1, \alpha_1) \cdot (c_2, \alpha_2) = \{\alpha_2(e_1) - \alpha_1(e_2)\}$ is onto. Let $(\beta, f) \in E^* \times E^{**} \approx E^* \times E$. We can take $e_1 = f$, $\alpha_1 = -\beta$; then (e_1, α_1) is mapped to (β, f) under $2\omega^b$. Conversely if ω^b is onto, then for $(\beta, f) \in E^* \times E^{**}$, there is (e_1, α_1) such that $f(\alpha_2) + \beta(e_2) = \alpha_2(e_1) - \alpha_1(e_2)$ for all e_2, α_2. Setting $e_2 = 0$ we see $f(\alpha_2) = \alpha_2(e_1)$, so $i: E \to E^{**}$ is onto. $\square$

Symplectic Forms induced by Metrics.

If $\langle, \rangle_x$ is a weak Riemannian metric on M, we have a smooth map $\varphi : TM \to T^*M$ defined by $\varphi(v_x)w_x = \langle v_x, w_x \rangle_x$, $x \in M$. If $\langle, \rangle$ is a (strong) Riemannian metric it follows from the implicit function theorem that φ is a diffeomorphism of TM onto T^*M. In any case, set $\Omega = \varphi^*(\omega)$ where ω is the canonical form on T^*M. Clearly Ω is exact since $\Omega = d(\varphi^*(\theta))$.

Proposition. (a) <u>If $\langle, \rangle_x$ is a weak metric, then Ω is a weak symplectic form. In a chart U for M we have</u>

$$\Omega(x,e)((e_1,e_2), (e_3,e_4)) = D_x\langle e,e_1\rangle_x e_3 - D_x\langle e,e_3\rangle_x e_1 + \langle e_4,e_1\rangle_x - \langle e_2,e_3\rangle$$

<u>where D_x denotes the derivative with respect to x.</u>

(b) <u>If $\langle,\rangle_x$ is a strong metric and M is modelled on a reflexive space, then Ω is a symplectic form.</u>

(c) $\Omega = d\theta$ <u>where, locally,</u> $\theta(x, e)(e_1, e_2) = -\langle e, e_1\rangle_x$.

<u>Note</u>. In the finite dimensional case, the formula for Ω becomes

$$\Omega = \Sigma\, g_{ij} dq^i \wedge d\dot{q}^i + \Sigma\, \frac{\partial g_{ij}}{\partial q^k} \dot{q}^i dq^j \wedge dq^k$$

where $q^1, \ldots, q^n, \dot{q}^1, \ldots, \dot{q}^n$ are coordinates for TM.

<u>Proof</u>. By definition of pull-back, $\Omega(x, e)((e_1, e_2), (e_3, e_4)) = \omega(x, e)(D\varphi_{(x,e)}(e_1, e_2), D\varphi_{(x,e)}(e_3, e_4))$. But clearly $D\varphi_{(x,e)}(e_1, e_2) = (e_1, D_x\langle e, \cdot\rangle_x e_1 + \langle e_2, \cdot\rangle_x)$, so the formula for Ω follows from that for ω. To check weak nondegeneracy, suppose $\Omega_{(x,e)}((e_1, e_2),(e_3, e_4)) = $ for all (e_3, e_4). Setting $e_3 = 0$ we find $\langle e_4, e_1\rangle_x = 0$ for all e_4, whence $e_1 = 0$. Then we obtain $\langle e_2, e_3\rangle_x = 0$, so $e_2 = 0$. Part (b) follows from the easy fact that the transform of a symplectic form by a diffeomorphism is still symplectic. □

The above result holds equally well for pseudo-Riemannian manifolds.

Note that if $M = H$ is a Hilbert space with the constant inner product, then ω is, on $H \times H$ which we may identify with $\mathcal{H}$ – the complexified Hilbert space, equal to the imaginary part of the inner product: Let $e = e_1 + ie_2$, $f = f_1 + if_2$. Then

$$\langle e, f \rangle = (\langle e_1, f_1 \rangle + \langle e_2, f_2 \rangle) - i(\langle e_1, f_2 \rangle - \langle e_2, f_1 \rangle)$$

so

$$\omega(e, f) = -\mathrm{Im}\langle e, f \rangle .$$

Canonical Transformations.

Let P, ω be a weak symplectic manifold; i.e. ω is a weak symplectic form on P. A (smooth) map $f : P \to P$ is called <u>canonical</u> or <u>symplectic</u> when $f^*\omega = \omega$. It follows that $f^*(\omega \wedge \ldots \wedge \omega) = \omega \wedge \ldots \wedge \omega$ (k times). If P is 2n dimensional, $\mu = \omega \wedge \ldots \wedge \omega$ (n times) is nowhere vanishing; by a computation one finds μ to be a multiple of the Lebesgue measure in canonical coordinates. We call μ the <u>phase volume</u> or the <u>Liouville form</u>. Thus a symplectic map preserves the phase volume, and is necessarily a local diffeomorphism.

We briefly discuss symplectic maps induced by maps on the base space of a cotangent bundle.

<u>Theorem. Let M be a manifold and</u> $f : M \to M$ <u>a diffeomorphism</u>, <u>define the lift of f by</u>

$$T^*f : T^*M \to T^*M ; \quad T^*f(\alpha_m)v = \alpha_m(Tf \cdot v) , \; v \in T^*_{f^{-1}(m)} M .$$

Then T^*f <u>is symplectic and in fact</u> $(T^*f)^*\theta = \theta$, <u>where θ is the canonical one form.</u> (We could, equally well consider diffeomorphisms from one manifold to another.)

Proof. By definition, $(T^*f)^*\theta(W) = \theta(TT^*f \cdot W) =$
$-T^*f(\alpha_m) \cdot (T\tau^*TT^*f \cdot W) = -T^*f(\alpha_m) \cdot (T(\tau^* \circ T^*f) \cdot W)$
$$= -\alpha_m \cdot (Tf \cdot T(\tau^* \circ T^*f) \cdot W)$$
$$= -\alpha_m \cdot (T(f \circ \tau^* \circ T^*f) \cdot W)$$
$$= -\alpha_m \cdot (T\tau^* \cdot W) = \theta(W)$$
since, by construction, $f \circ \tau^* \circ T^*f = \tau^*$. □

One can show conversely that any diffeomorphism of $P = T^*M$ which preserves θ is the lift of some diffeomorphism of M. But, on the other hand, there are many other symplectic maps of P which are not lifts.

Corollary. *Let M be a weak Riemannian manifold and Ω the corresponding weak symplectic form. Let $f : M \to M$ be a diffeomorphism which is an isometry:* $\langle v, w \rangle_x = \langle Tf \cdot v, Tf \cdot w \rangle_{f(x)}$. *Then $Tf : TM \to TM$ is symplectic.*

Proof. The result is immediate from the above and the fact that $T^*f \circ \varphi \circ Tf = \varphi$ where $\varphi : TM \to T^*M$ is as on p.34. □

Hamiltonian Vector Fields and Poisson Brackets.

Definition. Let P, ω be a weak symplectic manifold. A vector field $X : D \to TP$ with manifold domain D is called <u>Hamiltonian</u> if there is a C^1 function $H : D \to R$ such that

$$i_X \omega = dH$$

as 1-forms on D. We say X is <u>locally Hamiltonian</u> if $i_X\omega$ is closed.

We write $X = X_H$ because usually in examples one is given H and then one constructs the Hamiltonian vector field X_H.

Because ω is only weak, given $H : D \to \mathbb{R}$, X_H need not exist. Also, even if H is smooth on all of P, X_H will in general be defined only on a certain subset of P, but where it is defined, it is unique.

The condition $i_{X_H}\omega = dH$ reads

$$\omega_x(X_H(x), v) = dH(x) \cdot v,$$

$x \in D$, $v \in T_x D \subset T_x M$. From this we note that, necessarily, for each $x \in D$, $dH(x) : T_x D \to \mathbb{R}$ is extendable to a bounded linear functional on $T_x P$.

The relation $\omega(X_H, v) = dH \cdot v$ is the geometrical formulation of the same condition $X_H(\xi) = J \cdot \text{grad } H(\xi)$ with which we motivated the discussion.

<u>Some Properties of Hamiltonian Systems</u>.

We now give a couple of simple properties of Hamiltonian systems. The proofs are a bit more technical for densely defined vector fields so for purposes of these theorems we work with C^r vector fields.

Theorem. Let X_H be a Hamiltonian vector field on the symplectic manifold P, ω and let F_t be the flow of X_H. Then

(i) F_t is symplectic, $F_t^*\omega = \omega$

and (ii) energy is conserved; $H \circ F_t = H$.

Proof. (i) Since F_0 = identity, it suffices to show that $\frac{d}{dt} F_t^*\omega = 0$. But by lecture 1,

$$\frac{d}{dt} F_t^*\omega(x) = F_t^*(L_{X_H}\omega)(x)$$

$$= F_t^*[di_{X_H}\omega](x) + F_t^*[i_{X_H} d\omega](x)$$

The first term is zero because it is ddH and the second is zero because $d\omega = 0$.

(ii) By the chain rule,

$$\frac{d}{dt}(H \circ F_t)(x) = dH(F_t(x)) \cdot X_H(F_t(x))$$

$$= \omega_{F_t(x)}(X_H(F_t(x)), X_H(F_t(x)))$$

but this is zero in view of the skew symmetry of ω. □

An immediate corollary of (i) is <u>Liouville's theorem</u>: F_t <u>preserves the phase volume</u>. It seems likely that a version of Liouville's theorem holds in infinite dimensions as well. The phase volume would be a Wiener measure induced by the symplectic form.

More generally than (ii) one can show that for any function $f : P \to \mathbb{R}$,

$$\frac{d}{dt} f \circ F_t = \{f, H\} \circ F_t$$

where $\{f, g\} = \omega(X_f, X_g)$ is the Poisson bracket; in fact it is easy to see that

$$\{f, g\} = L_{X_g} f .$$

(Note that $F_t^* f = f \circ F_t$ for functions.)

The Wave Equation as a Hamiltonian System.

The wave equation for a function $u(x, t)$, $x \in R^n$, $t \in R$ is given by

$$\frac{\partial^2 u}{\partial t^2} = \Delta u + m^2 u , \quad m \geq 0$$

with u given at $t=0$, We consider

$$P = H^1(R^n) \times L_2(R^n)$$

where H^1 consists of functions in L_2 whose first derivatives are also in L_2. Let

$$D = H^2 \times H^1$$

and

$$X_H(u, \dot{u}) = (\dot{u}, \Delta u + m^2 u)$$

with symplectic form that associated with the L_2 metric

$$\omega((u, \dot{u}), (v, \dot{v})) = \int \dot{v} u - \int \dot{u} v .$$

(Recall that there is always an associated complex structure -- in this case that of $L_2(R, C)$; in fact there is also one making the flow of X_H unitary as in Cook [1], at least if $m > 0$) . Define

$$H(u, \dot{u}) = \frac{1}{2} \int \dot{u}^2 + \frac{1}{2} \int \|\nabla u\|^2 .$$

It is an easy verification (integration by parts) that X_H , ω and H are in the proper relation, so in this sense the wave equation is Hamiltonian.

That this equation has a flow on P follows from the hyperbolic version of the Hille-Yosida theorem stated in lecture 1.

The Schrodinger Equation.

Let $P = \mathcal{H}$ a complex Hilbert space with $\omega = \text{Im} \langle , \rangle$. Let H be a self adjoint operator with domain D and let

$$X_H(\varphi) = iH \cdot \varphi$$

and

$$H(\varphi) = \langle H\varphi, \varphi \rangle / 2 , \quad \varphi \in D .$$

Again it is easy to check that ω , X_H and H are in the correct relation.

In this sense X_H is Hamiltonian. Note that $\psi(t)$ is an integral curve of X_H if

$$\frac{1}{i}\frac{d\psi}{dt} = H\psi ,$$

the <u>abstract Schrodinger equation</u> of quantum mechanics.

That X_H has a flow is another case of the Hille-Yosida theorem called <u>Stone's theorem</u>; i.e. <u>if H is self adjoint, then iH generates a one parameter unitary group, denoted</u> e^{itH}.

We know from general principles that the flow e^{itH} will be symplectic. The additional structure needed for unitarity is exactly complex linearity.

We shall return to quantum mechanical systems in a later lecture.

We next turn our attention to geodesics and more generally to Lagrangian systems.

The Spray of a Metric.

Let M be a weak Riemannian manifold with metric $<,>_x$ on the tangent space T_xM. We now wish to define the <u>spray</u> S of the metric $<,>_x$. This should be a vector field on TM ; $S : TM \to T^2M$ whose integral curves project onto geodesics. Locally, if $(x, v) \in T_xM$, write $S(x, v) = ((x, v), (v, \gamma(x, v)))$. If M is finite dimensional, the geodesic spray is given by putting $\gamma^i(x, v) = -\Gamma^i_{jk}(x)v^jv^k$. In the general case, the correct definition for γ is

(1) $\quad <\gamma(x, v), w>_x \equiv \frac{1}{2}D_x<v, v>_x \cdot w - D_x<v, w>_x \cdot v$

where $D_x \langle v, v \rangle_x \cdot w$ means the derivative of $\langle v, v \rangle_x$ with respect to x in the direction of w. In the finite-dimensional case, the right hand side of (1) is given by

$$\frac{1}{2} \frac{\partial g_{ij}}{\partial x^k} v^i v^j w^k - \frac{\partial g_{ij}}{\partial x^k} v^i w^j v^k ,$$

which is the same as $-\Gamma^i_{jk} v^j v^k w_i$. So with this definition of γ, S is taken to be the spray. The verification that S is well-defined independent of the charts is not too difficult. Notice that γ is quadratic in v. One can also show that S is just the Hamiltonian vector field on TM associated with the kinetic energy $\frac{1}{2}\langle v, v \rangle$. This will actually be done below; cf. Abraham [2] and Chernoff-Marsden [1].

The point is that the definition of γ in (1) makes sense in the infinite as well as the finite dimensional case, whereas the usual definition of Γ^i_{jk} makes sense only in finite dimensions. This then gives us a way to deal with geodesics in infinite dimensional spaces.

Equations of Motion in a Potential.

Let $t \mapsto (x(t), v(t))$ be an integral curve of S. That is:

(2) $\qquad \dot{x}(t) = v(t) ; \quad \dot{v}(t) = \gamma(x(t), v(t))$.

These are the equations of motion in the absence of a potential. Now let $V : M \to R$ (the potential energy) be given. At each x, we have the differential of V, $dV(x) \in T_x^*M$, and we define $\text{grad } V(x)$ by:

(3) $$\langle \text{grad } V(x), w \rangle_x \equiv dV(x) \cdot w .$$

It is a definite assumption that $\text{grad } V$ exists, since the map $T_xM \to T_x^*M$ induced by the metric is not necessarily bijective.

The equation of motion in the potential field V is given by:

(4) $$\dot{x}(t) = v(t) ; \quad \dot{v}(t) = \gamma(x(t), v(t)) - \text{grad } V(x(t)) .$$

The total energy, kinetic plus potential, is given by $H(v_x) = \frac{1}{2}\|v_x\|^2 + V(x)$. It is actually true that the vector field X_H determined by H and the symplectic structure on TM induced by the metric is given by (4). This will be part of a more general derivation of Lagrange's equations below.

Lagrangian Systems.

We now want to generalize the idea of motion in a potential to that of a Lagrangian system; these are, however, still special types of Hamiltonian systems. See Abraham [2] for an alternative exposition of the finite dimensional case, and Marsden [1], and Chernoff-Marsden [1] for additional results.

We begin with a manifold M and a given function $L : TM \to R$ called the Lagrangian. In case of motion in a potential, one takes

$$L(v_x) = \frac{1}{2} \langle v_x, v_x \rangle - V(x)$$

which differs from the energy in that we use $-V$ rather than $+V$.

Now L defines a map, called the <u>fiber derivative</u>, $FL : TM \to T^*M$ as follows: let $v, w \in T_xM$. Then set

$$FL(v) \cdot w \equiv \frac{d}{dt} L(v + tw) \big|_{t=0}$$

That is, $FL(V) \cdot w$ is the derivative of L along the fiber in direction w.

In case of $L(v_x) = \frac{1}{2} \langle v_x, v_x \rangle_x - V(x)$, we see that $FL(v_x) \cdot w_x = \langle v_x, w_x \rangle_x$ so we recover the usual map of $TM \to T^*M$ associated with the bilinear form $\langle, \rangle_x$.

As we saw above, T^*M carries a canonical symplectic form ω. Using FL we obtain a closed two form ω_L on TM by

$$\omega_L = (FL)^* \omega .$$

In fact a straightforward computation yields the following local formula for ω_L: if M is modeled on a linear space E, so locally TM looks like $U \times E$ where $U \subset E$ is open, then $\omega_L(u, e)$ for $(u, e) \in U \times E$ is the skew symmetric bilinear form on $E \times E$ given by

$$\omega_L(u, e) \cdot ((e_1, e_2), (e_3, e_4)) = D_1(D_2L(u, e) \cdot e_3) \cdot e_1$$

$$- D_1(D_2L(u, e) \cdot e_3) \cdot e_1 + D_2D_2L(u, e) \cdot e_4 \cdot e_1 - D_2D_2L(u, e) \cdot e_2 \cdot e_3$$

where D_1, D_2 denote the indicated partial derivatives of L.

It is easy to see that ω_L is (weakly) nondegenerate if $D_2D_2L(u, e)$ is (weakly) nondegenerate. But we want to also allow degenerate cases for later purposes. In case of motion in a potential, nondegeneracy of ω_L amounts to nondegeneracy of the metric $<,>_x$. The __action__ of L is defined by $A : TB \to R$, $A(v) = FL(v) \cdot v$, and the __energy__ of L is $E = A - L$. In charts,

$$E(u, e) = D_2L(u, e) \cdot e - L(u, e)$$

and in finite dimensions it is the expression

$$E(q, \dot{q}) = \frac{\partial L}{\partial \dot{q}^i} \dot{q}^i - L(q, \dot{q}),$$

(summation convention!)

Now given L, we say that a vector field Z on TM is a __Lagrangian vector field or a Lagrangian system__ for L if the __Lagrangian condition__ holds:

$$\omega_L(v)(Z(v), w) = dE(v) \cdot w$$

for all $v \in T_bM$, and $w \in T_v(TM)$. Here, dE denotes the differential of E.

Below we shall see that for motion in a potential, this leads to the same equations of motion which we found above.

If ω_L were a weak symplectic form there would be at most one such Z. The fact that ω_L may be degenerate however means that Z is not uniquely determined by L so that there is some arbitrariness in what we may choose for Z. Also if ω_L is degenerate, Z may not even exist. If it does, we say that we can define <u>consistent equations of motion</u>. These ideas have been discussed in the finite dimensional case by Dirac [1] and Kunzle [1].

The dynamics is obtained by finding the integral curves of Z; that is the curves $v(t)$ such that $v(t) \in TM$ satisfies $(dv/dt)(t) = Z(v(t))$. From the Lagrangian condition it is trivial to check that energy is conserved even though L may be degenerate:

<u>Proposition.</u> <u>Let Z be a Lagrangian vector field for L and let $v(t) \in TM$ be an integral curve of Z. Then $E(v(t))$ is constant in t.</u>

<u>Proof.</u> By the chain rule,

$$\frac{d}{dt} E(v(t)) = dE(v(t)) \cdot v'(t) = dE(v(t)) \cdot Z(v(t))$$

$$= 2\omega_L(v(t))(Z(v(t)), Z(v(t))) = 0$$

by the skew symmetry of ω_L. □

We now want to generalize our previous local expression for the spray of a metric, and the equations of motion in the presence of a potential. In the general case the equations are called "Lagrange's equations".

Proposition. Let Z be a Lagrangian system for L and suppose Z is a second order equation (that is, in a chart $U \times E$ for TM, $Z(u, e) = (e, Z_2(u, e))$ for some map $Z_2 : U \times E \to E$). Then in the chart $U \times E$, an integral curve $(u(t), v(t)) \in U \times E$ of Z satisfies Lagrange's equations:

(1)
$$\begin{cases} \dfrac{du}{dt}(t) = v(t) \\ \dfrac{d}{dt}(D_2 L(u(t), v(t)) \cdot w = D_1 L(u(t), v(t)) \cdot w \end{cases}$$

for all $w \in E$. In case L is nondegenerate we have

(2) $\quad \dfrac{dv}{dt} = \{D_2 D_2 L(u, v)\}^{-1} \{D_1 L(u, v) - D_1 D_2 L(u, v) \cdot v\}$.

In case of motion in a potential, (2) reduces readily to the equations we found previously defining the spray and gradient.

Proof. From the definition of the energy E we have
$dE(u, e) \cdot (e_1, e_2) = D_1(D_2 L(u, e) \cdot e_1 + D_2 D_2 L(u, e) \cdot e \cdot e_2 - D_1 L(u, e) \cdot e_1$.
Locally we may write $Z(u, e) = (e, Y(u, e))$ as Z is a second order equation. Using the formula for ω_L, the condition on Y may be written, after a short computation:

$$D_1 L(u, e) \cdot e_1 = D_1(D_2 L(u, e) \cdot e_1) \cdot e + D_2(D_2 L(u, e) \cdot Y(u, e)) \cdot e_1$$
$$\text{for all } e_1 \in E.$$

This is the formula (2) above. Then, if $(u(t), v(t))$ is an integral curve of Z we obtain, using dots to denote time differentiation,

$$D_1 L(u, e) \cdot e_1 = D_1(D_2 L(u, \dot{u}) \cdot e_1 \cdot \dot{u} + D_2 D_2 L(u, \dot{u}) \cdot \ddot{u} \cdot e_1$$

$$= \frac{d}{dt} D_2 L(u, \dot{u}) \cdot e_1$$

by the chain rule. □

From these calculations one sees that if ω_L is nondegenerate Z is automatically a second order equation (cf. Abraham [2]). Also, the condition of being second order is intrinsic; Z is second order if $T\pi \circ Z$ = identity, where $\pi : TM \to M$ is the projection. See Abraham [2], or Lang [1].

Often L is obtained in the form

$$L(u, \dot{u}) = \int_Q \mathcal{L}(u, \frac{\partial u}{\partial x^k}, \dot{u}) d\mu$$

for a Lagrangian density $\mathcal{L}$ and μ some volume element on some manifold Q. Then M is a space of functions on Q or more generally sections of a vector bundle over Q. In this case, Lagrange's equations may be converted to the usual form of Lagrange's equations for a density $\mathcal{L}$. We shall see how this is done in a couple of special cases in later lectures. (See also Marsden [1]).

3. Elliptic Operators and Function Spaces.

In this lecture we shall discuss some of the basic spaces of functions which are used in analysis. In addition we shall discuss some of the fundamental properties of elliptic operators, first in the case of the Laplacian, and then in general. These results, especially the "splitting theorems" are of considerable use in proving certain subsets of the function spaces are actually submanifolds. This will find application in hydrodynamics and general relativity. Finally, we shall consider some elementary properties of the space of maps of one manifold to another.

We begin then with a discussion of the Sobolev spaces.

Sobolev spaces.

Let $\Omega \subset R^n$ be an open bounded set with C^∞ boundary. Let $\overline{\Omega}$ be the closure of Ω. Define $C^\infty(\Omega, R^n)$ to be the set of functions from Ω into R^n that can be extended* to a C^∞ function on some open set in R^n containing $\overline{\Omega}$. Let $C_0^\infty(\Omega, R^m) = \{f \in C^\infty(\Omega, R^m) \mid$ the support of f is contained in a compact subset of $\Omega\}$.

To describe the Sobolev spaces in an elementary fashion, we temporarily introduce some more notation. An n multi-index is

---

* This definition is the same as saying that the functions are C^∞ on the closed set $\overline{\Omega}$ (with difference quotients taken within $\overline{\Omega}$) by virtue of the Whitney extension theorem. See the appendix of Abraham-Robbin [1]. The same technique can be applied to Sobolev spaces; cf. the Calderon extension theorem below and Marsden [8].

an ordered set of n non-negative integers. If $k = (k_1, \ldots, k_n)$ is an n multi-index, then put $|k| = k_1 + k_2 + \ldots + k_n$. If $u \in C^\infty(\Omega, R^m)$, define $D^k u$ by the formula

$$D^k u = (\partial^{|k|} u / \partial x_1^{k_1} \ldots \partial x_n^{k_n})$$

and $D^0(u) = u$. For $u \in C^\infty(\Omega, R^m)$ (or $C_0^\infty(\Omega, R^n)$), define

$$\|u\|_s^2 = \int_\Omega \sum_{0 \leq |k| \leq s} |D^k u(x)|^2 dx .$$

Now $H^s(\Omega, R^m)$ (resp. $H_0^s(\Omega, R^m)$) is defined to be the completion of $C^\infty(\Omega, R^m)$ (resp. $C_0^\infty(\Omega, R^m)$) under the $\|\ \|_s$ norm. These H^s spaces are called the <u>Sobolev spaces</u>. Note that $H_0^0(\Omega, R^m) = H^0(\Omega, R^m) = L_2(\Omega, R^m) \supset H^s(\Omega, R^m)$; but for $s \geq 1$, $H_0^s(\Omega, R^m) \neq H^s(\Omega, R^m)$ as we shall see below.

There is another equivalent, and perhaps better, definition of the Sobolev norm. Let $d^k u$ be the kth total derivative of u so that $d^k u : \Omega \to L^k(R^n, R^m)$ where $L^k(R^n, R^m)$ denotes the k-linear maps on $\underbrace{R^n \times R^n \times \ldots \times R^n}_{k \text{ times}} \to R^m$ with the standard norm. Then if we set

$$|u|_s^2 = \int_\Omega \sum_{0 \leq k \leq s} \|d^k u(x)\|^2 dx$$

the $|\ |_s$ and $\|\ \|_s$ norms are equivalent. This is a simple exercise.

Also note that $H^s(\Omega, R^n)$ and $H_0^s(\Omega, R^n)$ are Hilbert spaces with the inner product

$$\langle u, v\rangle = \int_\Omega \sum_{0\le|k|\le s} D^k u(x)\cdot D^k v(x)\,dx .$$

Sobolev Theorem.

(a) <u>Let</u> $s > (n/2) + k$. <u>Then</u> $H^s(\Omega, R^m) \subset C^k(\Omega, R^m)$ <u>and the inclusion map is continuous (in fact is compact) when</u> $C^k(\Omega, R^m)$ <u>has the standard</u> C^k <u>topology, (the sup of the derivatives of order $\le k$).</u>

(b) <u>If</u> $s > (n/2)$ <u>then</u> $H^s(\Omega, R^m)$ <u>is a ring under pointwise multiplication of components. (This is often called the Schauder ring.)</u>

(c) <u>If</u> $s > \frac{1}{2}$ <u>and</u> $f \in H^s(\Omega, R^m)$ <u>then</u> $f|\partial\Omega \in H^{s-\frac{1}{2}}$.

(d) (<u>Calderon Extension Theorem</u>). <u>If</u> $f \in H^s(\Omega, R^m)$ <u>then f has an extension</u> $\tilde{f} \in H^s(R^n, R^m)$.

Regarding (c), see Palais [1] for a discussion of continuous Sobolev chains; i.e., the definition of H^s for s not an integer; basically one can use the Fourier transform or one can interpolate. (d) means that f can be extended across $\partial\Omega$ in an H^s way.

Differentiability properties at the boundary presents some technical problems but are very important in hydrodynamics. Thus it is important to distinguish H_0^s from H^s.

The proof of the Sobolev Theorem can be found in Nirenberg

[1] and Palais [1]; see also Sobolev [1].

For most of hydrodynamics we will need $s > (n/2)+1$. One of the outstanding problems in the field is determining to what extent we can relax this condition on s. For many problems, one would like to allow corners and discontinuities in such things as the density of the fluid or the velocity field. $L_p^k = W^{k,p}$ spaces are often useful for this.

H^s Spaces of Sections.

Let M be a compact manifold, possibly with boundary. Also, let E be a finite dimensional vector bundle over M. For example E may be the tangent bundle, or a tensor bundle over M. Let $\pi : E \to M$ be the canonical projection. The following fact is useful and is obvious from the definition of a vector bundle (see lecture 1).

Proposition. *Suppose for each* $x \in M$, *we have* $\pi^{-1}(x) \cong R^m$. *Then there is a finite open cover* $\{U_i\}$ *of* M *such that each* U_i *is a chart of* M *and* $\pi^{-1}(U_i) \cong U_i \times R^m$ *for each* i.

Such a cover is called __trivializing__. Recall that a __section__ of E is a map $h : M \to E$ such that $\pi \circ h = \mathrm{id}_M$. Informally, we define, for $s \geq 0$, $H^s(E)$ to be the set of sections of E whose derivatives up to order s are in L_2.

This makes sense since in view of the proposition, a section of E can locally be thought of as a map from R^n to R^m where n

is the dimension of M. Similarly, we can put a Hilbert structure on $H^s(E)$ by using a trivializing cover. However, since this Hilbert space structure depends on the choice of charts, the norm on $H^s(E)$ is not canonical, so we call $H^s(E)$ a __Hilbertible Space__ (ie., it is a space on which some complete inner product exists). To obtain a good norm on $H^s(E)$ one needs some additional structure such as a connection.

One has to check that the definition of $H^s(E)$ is independent of the trivialization and this can be done by virtue of compactness of M.

Of course the Sobolev theorems have analogues for $H^s(E)$. In particular if $s \geq 1$ it makes sense to restrict a section $h \in H^s(E)$ to ∂M. This is by part (c) of the Sobolev Theorem. Of course if $s > (n/2)$, h will be continuous and so this will be clear. For $s = 0$, we have $L_2(E)$ and restriction to ∂M does not make sense.

One defines $H^s_0(E)$ in a similar way. For $s > \frac{1}{2}$, when we restrict $h \in H^s_0(E)$ to ∂M, h will vanish, as will its derivatives to order $s - \frac{1}{2}$.

Much of the theory goes over for M noncompact, but we must specify a metric on M and a connection on E; further M must be complete and obey some curvature restriction such as sectional curvature bounded above; see Cantor [2].

Operations on Differential Forms.

Now, let M be a compact oriented Riemannian n-manifold without boundary.

As in Lecture 1, let Λ^k be the vector bundle over M whose fiber at $x \in M$ consists of k-linear skew-symmetric maps from $T_x M$, the tangent space to M at $x \in M$, to R. For each x, $\oplus_{k=0}^{n} \Lambda_x^k$ forms a graded algebra with the wedge product. Then $H^s(\Lambda^k)$ is a space of H^s differential k-forms. The exterior derivative d then is an operator:

$$d : H^{s+1}(\Lambda^k) \to H^s(\Lambda^{k+1}) .$$

It drops one degree of differentiability because d differentiates once; i.e., is a first order operator.

The star operator $* : H^s(\Lambda^k) \to H^s(\Lambda^{n-k})$ is given on Λ^k at $x \in M$ by

$$*(1) = \pm dx_1 \wedge \ldots \wedge dx_n , \quad *(dx_1 \wedge \ldots \wedge dx_n) = \pm 1$$

and

$$*(dx_1 \wedge \ldots \wedge dx_p) = \pm dx_{p+1} \wedge \ldots \wedge dx_n$$

where the "+" is taken if the $dx_1 \wedge \ldots \wedge dx_n$ is positively oriented and "-" otherwise, $x_1, \ldots, x_n$ form a coordinate system orthogonal at x, and $*$ is extended linearly as an operator

$\Lambda^k \to \Lambda^{n-k}$. Now if $\alpha \in H^s(\Lambda^k)$ then clearly $*\alpha \in H^s(\Lambda^{n-k})$, so $*$ can be taken as an operator from $H^s(\Lambda^k)$ to $H^s(\Lambda^{n-k})$.

The space Λ^k carries, at each point $x \in M$, an inner product. It is the usual business: the metric converts covariant tensors to contravariant ones (i.e., it raises or lowers indices) and then one contracts. If α_i, β_j are one forms, we have $\langle \alpha_1 \wedge \ldots \wedge \alpha_k, \beta_1 \wedge \ldots \wedge \beta_k \rangle = \det[\langle \alpha_i, \beta_j \rangle]$. It is not hard to check that if μ is the volume form on M then

$$\langle \alpha, \beta \rangle \mu = \alpha \wedge *\beta = \beta \wedge *\alpha .$$

Note that the inner product may be defined by the above formula. See Flanders [1] for more details on these matters. Define the operator $\delta : H^{s+1}(\Lambda^k) \to H^s(\Lambda^{k-1})$ by $\delta = (-1)^{n(k+1)+1} *d*$. There is an inner product on $H^0(\Lambda^k)$ (and hence on $H^s(\Lambda)$) given by

$$(\alpha, \beta) = \int_M \langle \alpha, \beta \rangle d\mu .$$

Proposition. <u>For</u> $\alpha \in H^s(\Lambda^k)$ <u>and</u> $\beta \in H^s(\Lambda^{k+1})$

$$(d\alpha, \beta) = (\alpha, \delta\beta) .$$

Proof. Note that $d(\alpha \wedge *\beta) = d\alpha \wedge *\beta + (-1)^k \alpha \wedge d*\beta$

$$= d\alpha \wedge *\beta - \alpha \wedge *\delta\beta ,$$

since $** = (-1)^{k(n-k)}$.

Since $\partial M = \emptyset$, by Stokes Theorem, we get

$$0 = \int_M d(\alpha \wedge *\beta)$$

$$= \int_M d\alpha \wedge *\beta - \int_M \alpha \wedge *\delta\beta$$

$$= (d\alpha, \beta) - (\alpha, \delta\beta) . \quad \square$$

Rephrasing, one says that d and δ are <u>adjoints</u> in the $(,)$ inner product.

The δ operator corresponds to the classical divergence operator. This is easily seen: let X be a vector field on M. Then because of the Riemannian structure X corresponds to a 1-form $\widetilde{X}$, where $\widetilde{X}(v) = \langle X, v \rangle$.

<u>Proposition.</u> $\operatorname{div}(X) = -\delta(\widetilde{X})$.

<u>Proof and Discussion.</u> Let $L_X \mu$ be the Lie derivative of μ with respect to X. Then by definition, $\operatorname{div}(X)\mu = L_X \mu$ (see Abraham [2]). We have the general formula

$$L_X \mu = d i_X(\mu) + i_X d(\mu) .$$

Now $d(\mu) = 0$ since μ is an n-form, so $L_X \mu = d(i_X \mu) = d(*\widetilde{X})$ (one easily checks that $i_X \mu = *\widetilde{X}$). Hence

$$\operatorname{div}(X) = \operatorname{div}(X)*\mu = *(\operatorname{div}(X)\mu) = *d*\widetilde{X} = -\delta\widetilde{X} ,$$

since for $k = 1$, $(-1)^{n(k+1)+1} = -1$. $\quad \square$

The <u>Laplace de Rham</u> operator is defined by $\Delta = \delta d + d\delta$. Note that $\Delta : H^s(\Lambda^k) \to H^{s-2}(\Lambda^k)$. If f is a real valued function on R^n, it is easy to check, using the above expressions for d, δ, that $\Delta(f) = -\nabla^2(f)$ where $\nabla^2 f = \text{div}(\text{grad } f)$ is the usual Laplacian. Note $\delta f = 0$ on functions.

<u>Proposition.</u> Let $\alpha \in H^s(\Lambda^k)$, <u>then</u> $\Delta\alpha = 0$ <u>iff</u>

$$d\alpha = 0 \text{ \underline{and} } \delta\alpha = 0.$$

<u>Proof.</u> It is obvious that if $d\alpha = 0$ and $\delta\alpha = 0$ then $\Delta\alpha = 0$. To show the converse, assume $\Delta\alpha = 0$. Then $0 = (\Delta\alpha, \alpha) = ((d\delta + \delta d)\alpha, \alpha) = (\delta\alpha, \delta\alpha) + (d\alpha, d\alpha)$, so the result follows. □

A form α for which $\Delta\alpha = 0$ is called <u>harmonic</u>.

<u>The Hodge decomposition theorem</u> (for $\partial M = \emptyset$).

<u>Theorem.</u> <u>Let</u> $\omega \in H^s(\Lambda^k)$. <u>Then there is</u> $\alpha \in H^{s+1}(\Lambda^{k-1})$, $\beta \in H^{s+1}(\Lambda^{k+1})$ <u>and</u> $\gamma \in C^\infty(\Lambda^k)$ <u>such that</u> $\omega = d\alpha + \delta\beta + \gamma$ <u>and</u> $\Delta(\gamma) = 0$. <u>Here</u> $C^\infty(\Lambda^k)$ <u>denotes the</u> C^∞ <u>sections of</u> Λ^k. <u>Furthermore</u> $d\alpha$, $\delta\beta$, <u>and</u> γ <u>are mutually</u> L_2 <u>orthogonal and so are uniquely determined.</u>

If $\mathcal{H}^k = \{\gamma \in C^\infty(\Lambda^k) | \Delta\gamma = 0\}$, then the above may be summarized by

$$H^s(\Lambda^k) = d(H^{s+1}(\Lambda^{k-1})) \oplus \delta H^{s+1}(\Lambda^{k+1}) \oplus \mathcal{H}^k.$$

The fact that the Harmonic forms $\mathcal{H}^k$ are all C^∞, follows from regularity theorems on the Laplacian. This fact is also called Weyl's lemma or, its generalization, Friedrich's theorem. We shall discuss this further below.

The Hodge theorem goes back to V. W. D. Hodge [1], in the 1930's. Substantial contributions have been made by many authors, leading up to the present theorem. See for example Weyl [1], and Morrey-Eells [1].

We can easily check that the spaces in the Hodge decomposition are orthogonal. For example

$$(d\alpha, \delta\beta) = (dd\alpha, \beta) = 0$$

since δ is the adjoint of d and $d^2 = 0$.

The basic idea behind the Hodge theorem can be abstracted as follows. We consider a linear operator T on a Hilbert space E with $T^2 = 0$. In our case $T = d$ and E is the L_2 forms. (We ignore the fact that T is only densely defined, etc.) Let T^* be the adjoint of T. Let $\mathcal{H} = \{x \in E | Tx = 0 \text{ and } T^*x = 0\}$. We assert

$$E = \overline{\text{Range } T} \oplus \overline{\text{Range } T^*} \oplus \mathcal{H}$$

which, apart from technical points on differentiability and so on is the essential content of the Hodge decomposition.

To see this, note, as before that the ranges of T and T^*

are orthogonal because

$$\langle Tx, T^*y\rangle = \langle T^2 x, y\rangle = 0 .$$

Let C be the orthogonal complement of $\overline{\text{Range } T} \oplus \overline{\text{Range } T^*}$. Certainly $\mathcal{H} \subset C$. But if $x \in C$,

$$\langle Ty, x\rangle = 0 \text{ for all } y \Rightarrow T^*x = 0 .$$

Similarly $Tx = 0$, so $C \subset \mathcal{H}$ and hence $C = \mathcal{H}$.

The complete proof of the theorem may be found in Morrey [1]. For more elementary expositions, also consult Flanders [1] and Warner [1].

An interesting consequence of this theorem is that $\mathcal{H}^k$ is isomorphic to the kth de Rham cohomology class (the clased k-forms mod the exact ones). This is clear since over M, each closed form ω may be written $\omega = d\alpha + \gamma$. (One can check that the $\delta\beta$ term drops out when ω is closed; indeed we get $0 = d\delta\beta$ so $(d\delta\beta, \beta) = 0$ or $(\delta\beta, \delta\beta) = 0$ so $\delta\beta = 0$.)

The Hodge theorem plays a fundamental role in incompressible hydrodynamics, as we shall see in lecture 4. It enables one to introduce the pressure for a given fluid state.

Below we shall generalize the Hodge theorem to yield some decomposition theorems for general elliptic operators (rather than the special case of the Laplacian). However, we first pause to discuss what happens if a boundary is present.

Hodge theory for manifolds with boundary.

This theory was worked out by Kodaira [1], Duff-Spencer [1], and Morrey [1]. (See Morrey [2], Chapter 7.) Differentiability across the boundary is very delicate, but important. The best possible results in this regard were worked out by Morrey.

Also note that d and δ may not be adjoints in this case, because boundary terms arise when we integrate by parts. Hence we must impose certain boundary conditions.

Let $\alpha \in H^s(\Lambda^k)$. Then α is __parallel__ or __tangent__ to ∂M if the normal part, $n\alpha = i^*(*\alpha) = 0$ where $i: \partial M \to M$ is the inclusion map. Analogously α is __perpendicular__ to ∂M if $t\alpha = i^*(\alpha) = 0$.

Let X be a vector field on M. Using the metric, we know when X is tangent or perpendicular to ∂M. X corresponds to the one-form $\tilde{X}$ and also to the $n-1$ form $i_X \mu$ (μ is, as usual, the volume form). Then X is tangent to ∂M if and only if $\tilde{X}$ is tangent to ∂M iff $i_X \mu$ is normal to ∂M. Similarly X is normal to ∂M iff $i_X \mu$ is tangent to ∂M. Set

$$H^s_t(\Lambda^k) = \{\alpha \in H^s(\Lambda^k) \mid \alpha \text{ is tangent to } \partial M\}$$

$$H^s_n(\Lambda^k) = \{\alpha \in H^s(\Lambda^k) \mid \alpha \text{ is perpendicular to } \partial M\}$$

and

$$\mathcal{H}^s(\Lambda^k) = \{\alpha \in H^s(\Lambda^k) \mid d\alpha = 0, \delta\alpha = 0\}.$$

The condition that $d\alpha = 0$ and $\delta\alpha = 0$ is now stronger than $\Delta\alpha = 0$. Following Kodaira [1], one calls elements of $\mathcal{H}^s$, <u>harmonic fields</u>.

<u>The Hodge Theorem.</u>

$$H^s(\Lambda^k) = d(H_t^{s+1}(\Lambda^{k-1})) \oplus \delta(H_n^{s+1}(\Lambda^{k+1})) \oplus \mathcal{H}^s(\Lambda^k) .$$

One can easily check from the formula

$$(d\alpha, \beta) = (\alpha, \delta\beta) + \int_{\partial M} \alpha \wedge *\beta$$

that the summands in this decomposition are orthogonal.

There are two other closely related decompositions that are of interest.

<u>Theorem.</u>

(a) $\quad H^s(\Lambda^k) = d(H^{s+1}(\Lambda^{k-1})) \oplus D_t^s$

where

$$D_t^s = \{\alpha \in H_t^s(\Lambda^k) \mid \delta\alpha = 0\}$$

and <u>dually</u>

(b) $\quad H^s(\Lambda^k) = \delta(H^{s+1}(\Lambda^{k+1})) + C_n^s$

where C_n^s <u>are the closed forms normal to</u> ∂M.

<u>Differential Operators and Their Symbols.</u>

Let E and F be vector bundles over M and let

$C^\infty(E)$, $H^s(E)$ denote the C^∞ and H^s sections of E as above. Assume M is Riemannian and the fibers of E and F have inner products.

A kth <u>order differential operator</u> is a linear map $D : C^\infty(E) \to C^\infty(F)$ such that if $f \in C^\infty(E)$ and f vanishes to kth order at $x \in M$, then $D(f)(x) = 0$. (Vanishing to kth order makes intrinsic sense independent of charts.)

Then in local charts D has the form

$$D(f) = \sum_{0 \leq |j| < k} a_j \frac{\partial^{|j|} f}{\partial x^{j_1} \ldots \partial x^{j_s}}$$

where $j = (j_1, \ldots, j_s)$ is a multi-index and a_j is a C^∞ function mapping E to F.

Now D has an <u>adjoint operator</u> D^* given in charts (with the standard Euclidean inner product on fibers) by

$$D^*(h) = \sum_{0 \leq |j| < k} (-1)^{|j|} \frac{1}{\rho} \frac{\partial^{|j|}}{\partial x^{j_1} \ldots \partial x^{j_s}} (\rho a_j^* h)$$

where $\rho \, dx^1 \wedge \ldots \wedge dx^n$ is the volume element and a_j^* is the transpose of a_j. The crucial property of D^* is

$$(g, D^*h) = (Dg, h)$$

where $(\,,\,)$ denotes the L_2 inner product, $g \in C_0^\infty(E)$, and $h \in C_0^\infty(F)$.

A kth order operator induces naturally a map

$$D : H^s(E) \to H^{s-k}(F).$$

For example we have the operators

$$d : H^s(\Lambda^k) \to H^{s-1}(\Lambda^{k+1})$$

$$\delta : H^s(\Lambda^k) \to H^{s-1}(\Lambda^{k-1})$$

and $\quad \Delta : H^s(\Lambda^k) \to H^{s-2}(\Lambda^k).$

The <u>symbol</u> of D assigns to each $\xi \in T^*_x M$, a linear map

$$\sigma_\xi : E_x \to F_x.$$

It is defined by

$$\sigma_\xi(e) = D(\frac{1}{k!}(g - g(x))^k f)(x)$$

where $g \in C^\infty(M, R)$, $dg(x) = \xi$ and $f \in C^\infty(E)$, $f(x) = e$. If there is danger of confusion we write $\sigma_\xi(D)$ to denote the dependence on D. By writing this out in coordinates one sees that σ_ξ is a polynomial expression in ξ of degree k obtained by substituting each ξ_{j_i} in place of a $\partial/\partial x^{j_i}$ in the highest order term. For example, if

$$D(f) = \Sigma \, g^{ij} \frac{\partial^2 f}{\partial x^i \partial x^j} + \text{(lower order terms)}$$

then

$$\sigma_\xi = \Sigma \, g^{ij} \xi_i \xi_j$$

(g^{ij} is for each ij a map of E_x to F_x). For real valued functions, the classical definition of an elliptic operator is that the above quadratic form be definite. This can be generalized as follows:

D is called <u>elliptic</u> if σ_ξ is an isomorphism for each $\xi \neq 0$.

We have now seen all three classical types of partial differential equations:

$$\text{elliptic: typified by } \Delta\alpha = \beta$$
$$\text{parabolic: typified by } \frac{\partial u}{\partial t} = \Delta u$$
$$\text{hyperbolic: typified by } \frac{\partial^2 u}{\partial t^2} = \Delta u .$$

To see that $\Delta : H^s(\Lambda^k) \to H^{s-2}(\Lambda^k)$ is elliptic one uses the facts that

(1) the symbol of d is $\sigma_\xi = \xi\wedge$

(2) the symbol of δ is $\sigma_\xi = i_\xi$

and (3) the symbol is multiplicative: $\sigma_\xi(D_1 \circ D_2) = \sigma_\xi(D_1) \circ \sigma_\xi(D_2)$.

<u>The Regularity Theorem and Splitting Theorems.</u>

<u>Theorem.</u> <u>Let</u> M <u>be compact without boundary.</u> <u>Let</u> D <u>be elliptic of order</u> k. <u>Let</u> $f \in L_2(E)$ <u>and suppose</u> $D(f) \in H^s(F)$. <u>Then</u> $f \in H^{s+k}(E)$.

One can allow boundaries if the appropriate boundary conditions are used. See Nirenberg [1]. As a special case of this theorem we get Weyl's lemma: $\Delta f = 0 \Rightarrow f$ is C^∞.

The proof of the theorem is too intricate to go into here; see Palais [1] or Yosida [1]. It is important to note that this sort of result is certainly false if we use C^k spaces, although Holder spaces $C^{k+\alpha}$, $0 < \alpha < 1$ would be suitable.

<u>Theorem</u>. (Fredholm Alternative)   <u>Let D be as above. Then</u>

$$H^s(F) = D(H^{s+k}(E)) \oplus \ker D^*$$

$(D^* : H^s(F) \to H^{s-k}(E))$. <u>Indeed this holds true if we merely assume that either D or D^* has injective symbol.</u>

The proof of this leans heavily on the regularity theorem. The main technical point is to show that $D(H^{s+k})$ is closed. (One uses the fact that $\|f\|_{s+k} \leq \text{const}(\|f\|_s + \|Df\|_s)$, for D elliptic.) Then one shows that the L_2 orthogonal complement of $D(H^{s+k})$ is in $\ker D^*$, just as in the Hodge argument. This yields an L_2 splitting and we get an H^s splitting via regularity. The splitting in case D has injective symbol relies on the fact that D^*D is, in this case, elliptic. One could use, e.g.: $D = d$ to get the Hodge theorem. For details on this, see Berger-Ebin [1].

In later applications (see lectures 4 and 10) we will use this result in the following way. Certain sets in which we are interested will be defined by constraints $f(x) = 0$. The relation $v \mapsto T_x f \cdot v$

will be a differential operator. To show it is surjective (and hence $f^{-1}(0)$ is a submanifold) we can show $(T_x f)^*$ is injective with injective symbol. For then $\ker(T_x f)^* = 0$, so $T_x f$ itself will be onto.

Manifolds of Maps.

History.

The basic idea was first laid down by Eells [1] in 1958. He constructed a smooth manifold out of the continuous maps between two manifolds. In 1961, Smale and Abraham worked out the more general case of C^k mappings. Their notes are pretty much unavailable, but the 1966 survey article by Eells [2] is a good reference. The H^s case is found in a 1967 article by Elliasson [1]. This is also found in Palais [4] where it is done in the more general context of fiber bundles.

Making the manifold out of the C^k diffeomorphism group on a compact manifold without boundary was done independently by Abraham (see Eells [2]) and Leslie [1] around 1966. The H^s case is found in a paper by Ebin [1] and one by Omori [1] around 1968. Ebin also showed that the volume preserving diffeomorphisms form a manifold. Finally Ebin-Marsden [1] worked out the manifold structure for the H^s diffeomorphisms, the symplectic and volume preserving diffeomorphisms for a compact manifold with smooth boundary.

Other papers on manifolds of maps include those of Saber [1],

Leslie [2, 3], Omori [2], Gordon [1], Penot [2, 3], and Graff [1].
Some further references are given below.

Local Structure.

Let M and N be compact manifolds and assume N is without boundary. Let n be the dimension of M, and ℓ the dimension of N. Say $f \in H^s(M, N)$ if for any $m \in M$ and any chart (U, φ) containing m and any chart (V, ψ) at $f(m)$ in N, the map $\psi \circ f \circ \varphi^{-1} : \varphi(U) \to R^\ell$ is in $H^s(\varphi(U), R^\ell)$. This can be shown to be a well defined notion, independent of charts for $s > (n/2)$. The basic fact one needs is that by the Sobolev Theorem we have $H^s(M, N) \subset C^0(M, N)$. Things are not as nice, however, for $s < (n/2)$. It is possible for a map to have a (derivative) singularity which is L_2 integrable in one coordinate system on N and not be integrable in another. So for $s < (n/2)$, $H^s(M, N)$ cannot be defined invariantly. Hence, from now on we assume $s > (n/2)$.

In order to find charts in $H^s(M, N)$ we first need to determine the appropriate modeling space. Let $f \in H^s(M, N)$. The modeling space, should it exist, must be isomorphic to $T_f H^s(M, N)$, whatever that is. So a way to begin is to find a plausible candidate for $T_f H^s(M, N)$. If P is any manifold and $p \in P$ then $T_p P$ can be constructed by considering any smooth curve c in P such that $c(0) = p$; then $c'(0) \in T_p P$ (see lecture 1).

With this in mind, let us consider a curve $c_f :]{-1}, 1[\to H^s(M, N)$

such that $c_f(0) = f$. Now if $m \in M$, then the function $t \mapsto c_f(t)(m)$ is a curve in N (i.e., for each $t \in\,]-1, 1[$, $c_f(t) \in H^s(M, N)$ and therefore $c_f(t) : M \to N$.) Now $c_f(0)(m) = f(m)$, so the derivative of this curve at 0, $(d/dt)c_f(t)(m)|_{t=0}$ is an element of $T_{f(m)}N$. So the map $m \mapsto (d/dt)c_f(t)(m)|_{t=0}$ maps M to TN and covers f, i.e., if $\pi_N : TN \to N$ is the canonical projection, this diagram commutes:

$$c'_f(0) \nearrow \begin{array}{c} TN \\ \downarrow \pi_N \\ \end{array} \qquad c'_f(0) = \frac{d}{dt}c_f(t)\Big|_{t=0}$$
$$M \xrightarrow{f} N$$

where

$$c'_f(0)(m) = \frac{d}{dt}c_f(t)(m)\Big|_{t=0} \,.$$

Making the identification

$$(\frac{d}{dt}c_f(t)|_{t=0})(m) = \frac{d}{dt}c_f(t)(m)\Big|_{t=0} \,,$$

$c'_f(0)$ is a good candidate for the tangent to c_f at f.

With the above motivation, let us define

$$T_f H^s(M, N) = \{X \in H^s(M, TN) \mid \pi_N \circ X = f\} \,.$$

Note this is a linear space, for if V_f and X_f are in $T_f H^s(M, N)$, we can define $aV_f + X_f$ ($a \in \mathbb{R}$) as the map $m \mapsto aV_f(m) + X_f(m)$ where

$V_f(m)$ and $X_f(m)$ are in $T_{f(m)}N$. It is this space which we use as a model for $H^s(M, N)$ near f.

To show this we need the map $\exp_N : T_pN \to N$ for $p \in N$. Recall that if $v_p \in T_pN$ there is a unique geodesic σ_{v_p} through p whose tangent vector at p is v_p. Then $\exp_p(v_p) = \sigma_{v_p}(1)$. In general $\exp_v$ is a diffeomorphism from some neighborhood of 0 in T_pN onto a neighborhood p in N. However, since N is compact and without boundary, it is geodesicly complete and hence $\exp_v$ is defined on all of T_pN. This map can be extended to a map $\exp : TN \to N$ such that if $v_p \in TN$ then $\exp(v_p) = \exp_p(v_p)$. With this map we define the map $\overline{\exp}_f : T_f H^s(M, N) \to H^s(M, N)$

$$X \mapsto \exp \circ X.$$

We assert that $\overline{\exp}_f$ maps the linear space $T_f H^s(M, N)$ <u>onto</u> a neighborhood of f in $H^s(M, N)$ taking 0 to f and hence is a candidate for a chart in $H^s(M, N)$. It should be remarked that in spite of the use of the map $\exp$, the structure is independent of the metric on N. The assertion is easy to check in case things are C^∞ or C^s, by using standard properties of $\exp$; Milnor [1].

For the H^s case and to show that the change of charts is well defined (i.e., maps into the right spaces) and is smooth, one needs the following lemma.

<u>Local ω-Lemma.</u> (Left Composition of Maps). <u>Let</u> U <u>be a bounded</u>

open set in R^p, and $h : R^n \to R^m$ be C^∞. Then $\omega_h : H^s(U, R^n) \to H^s(U, R^m)$ defined by $\omega_h(f) = h \circ f$ is a C^∞ map.

This conclusion is not true if h is merely an H^s or C^s map. The problem can be seen in this way. If M and N are manifolds and $g : M \to N$ is C^1 then for $p \in M$ and $v_p \in T_p M$, we have $T_p g : T_p M \to T_{g(p)} N$, which is determined in the usual way: let $c :]-1, 1[\in M$ be a curve such that $c(0) = p$ and $c'(0) = v_p$. Then $T_p g(v_p) = (d/dt) g(c(t))\big|_{t=0}$. Applying this procedure to ω_h, and using the chain rule, we find for $X \in T_f H^s(U, R^n)$ that the tangent of ω_h is the map $T_f \omega_h : X \mapsto Th \circ X$. But since Th is only H^{s-1}, $Th \circ X$ is, at best, in $H^{s-1}(U, R^m)$ and $T\omega_h$ does <u>not</u> map into the tangent space of $H^s(U, R^m)$ at $\omega_h(f)$.

This necessity of differentiating h is a crucial difference between composition on the left and composition on the right.

The exact proof of the ω-lemma may be found in Ebin [1] and the other references above. In fact, the result essentially goes back to Sobolev [1] p. 223. See also Marcus-Mizel [1], and Bourguinon - Brezis [1].

Using the ω-lemma, it is now routine to check that $\exp_f$ yields smooth charts on $H^s(M, N)$. For other methods of obtaining charts, see Palais [4], Penot [3] and Krikorian [1].

4. The Motion of an Incompressible Fluid.

This lecture is concerned with some fundamental properties of perfect fluids. We shall begin with some motivation and an intuitive outline of the results. Then we shall fill in a number of the gaps. The results of this section lean on work of Arnold [1] and Ebin-Marsden [1]. They are primarily concerned with interpreting the equations as a Hamiltonian system and with the associated existence theory. These go hand in hand, and as a bonus, when one regards the equations from the Hamiltonian point of view the existence theory unexpectedly becomes easier. The difference is essentially that between "Eulerian" and "Lagrangian (following the fluid)" coordinates, as we hope to explain. A similar bit of analysis can be made for elasticity.*

Basic Ideas in Hydrodynamics.

Throughout, let M be a fixed compact, oriented, Riemannian, n-manifold, possibly with a C^∞ boundary. Intuitively, M is the space in which the fluid moves. For example, M might be the unit ball in R^3. As an aside, for the general theory there seems to be no particular advantage of assuming M is open in R^n. This is because the spaces of mappings of M to M that we will shortly discuss are still very nonlinear.

A <u>diffeomorphism</u> on M is a C^∞ bijective map $\eta : M \to M$ such that η^{-1} is also C^∞.

---
*This remark is based on some recent joint work with T. Hughes.

We let $\mathcal{D} = \{$orientation preserving diffeomorphisms on $M\}$.

If the Riemannian structure is given locally by $g_{ij} : M \to R$, then the <u>volume element</u> μ on M is the n-form which, in a (positively oriented) coordinate chart, is given by

$$\mu = \sqrt{\det(g_{ij})}\, dx^1 \wedge \ldots \wedge dx^n$$

or, intrinsically, $\mu(v_1, \ldots, v_n) = \sqrt{\det\langle v_i, v_j \rangle}$ for $v_1, \ldots, v_n$ oriented tangent vectors. We say a diffeomorphism η is <u>volume preserving</u> if $\eta^*\mu = \mu$. The condition $\eta^*\mu = \mu$ means that the Jacobian of η is one.

By the change of variables formula, it follows that a diffeomorphism η is volume preserving if and only if for every measurable set $A \subset M$, $\mu(A) = \mu(\eta(A))$. Here we also use μ to stand for the measure defined by μ (cf. Abraham [2], §12).

Set $\mathcal{D}_\mu = \{\eta \in \mathcal{D} \mid \eta$ is volume preserving$\}$.

For technical reasons it will be convenient to enlarge $\mathcal{D}$ and $\mathcal{D}_\mu$ to slightly larger spaces. Namely let $\mathcal{D}^s$ (resp. $\mathcal{D}_\mu^s$) be the completion of $\mathcal{D}$, (resp. $\mathcal{D}_\mu$) under the Sobolev H^s topology; this will be discussed in detail later.

At least in the beginning, we will be discussing perfect fluids; i.e., fluids which are nonviscous, homogeneous and incompressible. We also ignore external forces for simplicity.

Consider, then, our manifold M whose points are supposed to represent the fluid particles at $t = 0$. Let us look at the fluid moving in M. As t increases, call $\eta_t(m)$ the curve followed by the fluid particle which is initially at $m \in M$. For fixed t, each η_t will be a diffeomorphism of M. In fact, since the fluid is incompressible, we have $\eta_t \in \mathcal{D}_\mu$. The function $t \mapsto \eta_t$ is thus a curve in $\mathcal{D}_\mu$ (they are easily seen to be orientation preserving since they are connected to η_0, the identity function on M). Note that if M has a fixed boundary the flow will be parallel to ∂M.

The motions of a perfect fluid are governed by the Euler equations which are as follows

(Euler equations) $\begin{cases} \dfrac{\partial v_t}{\partial t} + \nabla_{v_t} v_t = -\operatorname{grad} p_t \\[2pt] \operatorname{div} v_t = 0 \\[2pt] v_t \text{ is tangent to } \partial M. \end{cases}$

In these equations, $\nabla_{v_t} v_t$ is the covariant derivation and its ith component is given in a coordinate chart by

$$(\nabla_{v_t} v_t)^i = \sum_j v_t^j \frac{\partial v_t^i}{\partial x^j} + \sum_{j,k} \Gamma^i_{jk} v_t^j v_t^k$$

and $p_t = p(t)$ is some (unknown) real valued function on M called the <u>pressure</u>.

In the case of Euclidean space, each $\Gamma^i_{jk} = 0$ and then we get, using vector analysis notation

$$\nabla_v v = (\vec{v} \cdot \nabla) v .$$

Note. We shall always use a subscripted variable to denote that the variable is held fixed, as in v_t. It will never denote differentiation.

The physical derivation of these equations is quite simple in R^n. We use Newton's Law $F = ma$. We can ignore the mass because of homogeneity (i.e., constant mass density) and we are assuming there are no external forces, so the only forces result from the internal pressure. We wish to deal with conservative force fields and therefore one assumes these internal forces arise as the gradient of a real valued function, the pressure. So we have

$$\text{acceleration} = -\text{grad } p_t .$$

Clearly the acceleration is given by

$$a = \lim_{\Delta t \to 0} \frac{v(t + \Delta t, x(t + \Delta t)) - v(t, x(t))}{\Delta t}$$

$$= \frac{\partial v}{\partial t} + \sum_i \frac{\partial v}{\partial x^i} \frac{\partial x^i}{\partial t} = \frac{\partial v}{\partial t} + \sum_i \frac{\partial v}{\partial x^i} v^i .$$

Here we have just used the chain rule. This gives us the correct equation for $\partial v/\partial t$. Now div $v = 0$ is the same as assuming η_t is volume preserving, and v parallel to ∂M just corresponds to particles not moving across ∂M.

As the Euler equations stand, they are not Hamiltonian. In fact the way they are written, their form as an evolution equation is not manifest. To rectify the latter problem, we use:

Theorem. <u>Let X be an (H^s) vector field on M. Then X can be uniquely written</u>

$$X = Y + \text{grad } p$$

<u>for Y an (H^s) divergence free vector field parallel to ∂M and p an (H^{s+1}) function.</u>* <u>Here $s \geq 0$. We let $P(X) = Y$ and call P the projection onto the divergence free part.</u>

This follows directly from the Hodge decomposition applied to the corresponding one forms discussed in lecture 3.

Let E denote the space of all divergence free vector fields on M which are parallel to ∂M. Define $T : E \to E$ by $T(v) = -P(\nabla_v v)$. Note that

$$-P(\nabla_v v) = -(\nabla_v v - \text{grad } p) = -\nabla_v v + \text{grad } p$$

and therefore we can rewrite the Euler equation (modulo a trivial sign convention on p) as a differential equation on the linear space E:

$$\frac{\partial v_t}{\partial t} = T(v_t) \quad v_0 \text{ is given.}$$

---
* In case M were non-compact, e.g. $M = R^3$, p would only be locally H^{s+1}.

Notice that T maps H^s to H^{s-1} and so, as discussed in lecture one, the usual existence and uniqueness theorem doesn't apply. It is possible to use other methods however, such as the Nash-Moser technique and Galaerkin methods.

Observe that the equation is non-local. This is due to the non-local operator P; i.e., given X in a neighborhood of x_0, it is <u>not</u> possible to compute $P(X)$ at x_0. Rather P is an integral operator. Indeed, from

$$X = Y + \text{grad } p ,$$
$$\delta X = \Delta p$$
$$\text{so } p = \Delta^{-1} \delta X$$

and Δ^{-1} is an integral operator involving convolution with a suitable green's function.

Summary of the Main Results.

Before getting down to some technical details we would like to present the punch line. To do this we need to state a few facts proven below.

As above, let $\mathcal{D}$ denote all C^∞ diffeomorphisms of M, $\eta : M \to M$. One can show that $\mathcal{D}$ is (in a certain sense) a smooth manifold modelled on a Fréchet space and it is a "Lie group" in that the group operations of composition and inversion are smooth.

The tangent space to $\mathcal{D}$ at the identity, $T_e \mathcal{D}$ consists of all vector fields on M. This is as in lecture 3: Indeed, if $\eta(t) \in \mathcal{D}$ is a curve, $\frac{d}{dt} \eta(t)(m)\big|_{t=0}$ represents a vector field on M if $\eta(0)(m) = m$. Generally, $T_\eta \mathcal{D} = \{X : M \to TM | \pi \circ X = \eta\}$ where $\pi : TM \to M$ is the projection.

Also as above, we let $\mathcal{D}_\mu = \{\eta \in \mathcal{D} | \eta \text{ is volume preserving}\}$. Then $\mathcal{D}_\mu$ is also a "Lie group" and

$$T_e \mathcal{D}_\mu = \{X \in T_e \mathcal{D} | \text{divergence of } X = \text{div } x = 0\}.$$

If M has boundary we must always add the condition that X is parallel to the boundary.

Now put a metric on $\mathcal{D}$ and hence $\mathcal{D}_\mu$ by

$$(X, Y) = \int_M <X(m), Y(m)>_{\eta(m)} d\mu(m)$$

for $X, Y \in T_\eta \mathcal{D}$. It is easy to see that $(\,,\,)$ is right invariant on $\mathcal{D}_\mu$. This metric corresponds exactly to the total kinetic energy of the fluid:

$$\text{Energy} = \frac{1}{2} \int_M \|v\|^2 d\mu$$

where v is the velocity field of the fluid:

$$v_t(\eta_t(m)) = \frac{d}{ds} \eta_s(m)\big|_{s=t}$$

Given a time dependent vector field $v(t, x)$ satisfying the

Euler equations, we can construct its flow η_t ; it is the solution to

$$\begin{cases} \frac{d}{dt} \eta_t(m) = v_t(\eta_t(m)) \\ \eta_0(m) = m \end{cases}$$

and of course conversely, given η_t we can obtain v_t.

The first important fact is the following:

Theorem. (Arnold). A time dependent vector field $v(t, x)$ on M satisfies the Euler equations

$$\Leftrightarrow \text{ its flow } \eta_t \text{ is a geodesic in } \mathcal{D}_\mu .$$

The second one is:

Theorem. (Ebin-Marsden). The spray governing the geodesics on $\mathcal{D}_\mu$, $Z : T\mathcal{D}_\mu \to T^2\mathcal{D}_\mu$ is a C^∞ map in H^s : $Z : T\mathcal{D}_\mu^s \to T^2\mathcal{D}_\mu^s$. Hence the standard existence and uniqueness theorem can be used.

The first result is analogous to the way in which one can describe the motion of a rigid body either by looking at its velocity vector in Eulerian (space) coordinates or as a geodesic in the Lie group SO(3) (body coordinates). In fact one can proceed in general to describe Hamiltonian systems on Lie groups in general of which hydrodynamics and the rigid body are special cases (see Arnold [1], Marsden-Abraham [1] and Iacob [1]).

The second result should be surprizing in view of our

previous discussion that the standard existence and uniqueness theorem could <u>not</u> be used in Eulerian coordinates.

We would now like to try to give the essence of this idea. The key thing is that in Lagrangian coordinates, the equations change their character completely. Suppose then that

$$\frac{\partial v}{\partial t} + (v \cdot \vec{\nabla})v = -\text{grad } p$$

on R^3. We let η_t be the flow of v and look at the new variables η_t, $X = v_t \circ \eta_t$ instead of v itself. Now

$$\frac{\partial X}{\partial t} = \frac{\partial v}{\partial t} \circ \eta_t + \sum \frac{\partial v_t}{\partial x^i} \frac{\partial \eta_t^i}{\partial t}$$

$$= \frac{\partial v}{\partial t} \circ \eta_t + \sum \left(\frac{\partial v_t}{\partial x^i} \cdot v_t^i \right) \circ \eta_t = -\text{grad } p$$

since v satisfies $(\partial v/\partial t) + (v \cdot \vec{\nabla})v = -\text{grad } p$.

In order for the spray Z to be smooth, the map $(\eta, X) \mapsto (\frac{\partial \eta}{\partial t}, \frac{\partial X}{\partial t})$ has to at least map H^s to H^s. Now grad p is the gradient part of $(v \cdot \vec{\nabla})v$, so it is not completely obvious that grad p is H^s if v is H^s. However we can see it by a simple calculation: Indeed take the divergence of

$$\frac{\partial v}{\partial t} + \sum v^i \frac{\partial v}{\partial x^i} = -\text{grad } p$$

to get

$$-\Delta p = \frac{\partial}{\partial x^j}(v^i \frac{\partial v^j}{\partial x^i}) = \frac{\partial v^i}{\partial x^j}\frac{\partial v^j}{\partial x^i}$$

since div v = 0 . Thus if v is H^s , grad p will be H^s as well (regularity of the Laplace operator).

By combining the previous two theorems with the existence and uniqueness theorem, we obtain the following.

Corollary. Let $s > \frac{n}{2} + 1$ (n = dimension of M) , and v_0 a divergence free H^s vector field parallel to ∂M . Then there is a unique H^s v_t equalling v_0 at t = 0 which satisfies the Euler equations (that is, there is a p(x, t) such that the Euler equations hold), defined for $-\epsilon < t < \epsilon$ for some $\epsilon > 0$. If v_0 is C^∞ , so is v_t .

Recently Bourguignon and Brezis [1] have obtained these results in a more classical way without using infinite dimensional manifolds. The same results are also obtained in the spaces $W^{s,p}$, $s > \frac{n}{p} + 1$.

The question naturally arises if we can infinitely extend the solutions in the corollary. Such solutions would be called global.

Theorem. (Wolibner (1933), Judovich (1964), Kato (1967)). If dim M = 2 , the solutions in theorem 2 can be indefinitely extended for all $t \in R$ (and remain smooth).

The problem is open if dim M = 3 .

The problem is also open, in general, if we consider the equations with viscosity. This leads us to a hamiltonian system with a dissipative term.

Navier-Stokes equations
$$\begin{cases} \frac{\partial v}{\partial t} - \nu \Delta v + \nabla_v v = -\text{grad } p \quad (+ \text{ forces}) \\ \text{div } v = 0 \\ v = 0 \text{ on } \partial M \quad \text{(note the change in boundary conditions)} \end{cases}$$

The term $\nu \Delta v$ is an approximation to viscous forces in the fluid which tend to slow the fluid down. Thus the chances for a global solution are increased.

For the Euler equations it is known (see Marsden-Ebin-Fischer [1]) that if the C^1 norm of v is bounded on an interval $[0, T[$, then the solution can be extended beyond T. Thus one gets global solutions if an a priori bound is known on the C^1 norm. One can do better for the Navier-Stokes equations:

<u>Theorem.</u> (Leray [3]). <u>Let v_t be a solution to the Navier-Stokes equations</u>, dim $M = 3$. <u>Suppose one has an a priori bound on the spatial L_p norm of v_t on finite t-intervals, where $p > 3$. Then the solution can be infinitely extended to $[0, \infty[$ as a smooth solution.</u>

One can also show that one has global solutions if the initial data is sufficiently small (Ladyzhenskaya [2]) and for fixed but perhaps large initial data the time of existence is of the order of ν. It is really the case of "large" initial data which is of interest and for these, Leray's theorem gives a criterion which is necessary and sufficient, but is not too easy to verify (see remarks below).

In the next lecture we shall discuss a method, using "Chorin's formula", which gives a fundamental improvement on the time of existence for general initial data.

These difficulties with global solutions bear on the nature of turbulence. (See the next lecture for further discussion.) Indeed Leray believed that it is possible for solutions to become non-smooth and non-unique after some time interval $[0, T]$, at which time they turn into weak, or Hopf solutions and this was supposed to represent turbulence.

Nowadays, the opposite point of view prevails, but it is not yet completely settled. In other words, we now believe that turbulence represents very complicated, but still smooth solutions to the equations.

But the situation is very delicate and one must be careful. For example a law of Kolmogorov,* experimentally verified for turbulent flows, when translated into norms indicates that one has an a priori bound on the L_p norm of v_t for $p < 3$! This just misses the critical value of $p = 3$, but refinements of this may be able to raise the value of p above 3.

---

* We refer to the $-5/3$ law; see Landau-Lifschitz (1). The experimental verification is not conclusive and is also consistent with other possible laws.

There are other reasons for this view and we shall discuss them below in lecture 5. Briefly, turbulence is believed to be a result of successive losses of stability (rather than smoothness).

From Arnold's theorem we can rephrase the problem of extending solutions of the Euler equations as follows:

<u>Problem</u>. Let M be a (compact) 3-manifold. Then is $\mathcal{D}_\mu^s$ geodesically complete? (That is, do geodesics exist for all time? From Wolibner's result, the answer is yes, if $\dim M = 2$.) *

The following simple lemma bears on the problem (the lemma is standard); see Wolf [1], p. 89 and lecture 6 below.

<u>Lemma</u>. <u>Let</u> G <u>be a finite dimensional Lie group with a right invariant riemannian metric. Then</u> G <u>is geodesically complete</u>.

The lemma also holds if G is a "Hilbert group-manifold", but unfortunately, it does not apply to our problem because the topology of our metric (recall it gives the L^2 norm) does not coincide with the topology on $\mathcal{D}_\mu^s$. If the requirement $\operatorname{div} v = 0$ were dropped, the result is definitely false -- this is the phenomenon of shock waves in compressible flow. (For example the solution of $(\partial u/\partial t) + u(\partial u/\partial x) = 0$ in one dimension is $u(t, x) = u_0(y)$ where $x = y + tu_0(y)$. One can see as soon as $x \mapsto y$ becomes non-invertible, that derivatives of u blow up.)

---

* At present the most reasonable sounding conjecture for this problem is "no" because of "vortex sheets" but "yes" for the Navier-Stokes equations for which vortex sheets are impossible by Leray's theorem.

Kelvin Circulation Theorem.

This is a standard classical theorem of hydrodynamics that is very easy to prove in our context. It says the amount of circulation about any closed loop is constant in time.

Kelvin Circulation Theorem. *Let* M *be a manifold and* $\ell \subset M$ *a smooth closed loop i.e., a compact one manifold. Let* u_t *be a solution to the Euler Equations on* M *and* $\ell(t)$ *be the image of* ℓ *at time* t *when each particle moves under the flow* η_t *of* u_t *i.e.,* $\ell(t) = \eta_t(\ell)$. *Then*

$$\frac{d}{dt} \int_{\ell(t)} \tilde{u}_t = 0 \quad (\tilde{u}_t \text{ is the one form dual to } u_t).$$

Proof. We have the identity $L_u \tilde{u} = \tilde{\nabla_u u} + \tfrac{1}{2} d\langle u, u \rangle$, valid for any vector field u on the manifold M. We leave the verification as an exercise.

Then, identifying the differential forms with their dual vector fields, we find $P(L_u \tilde{u}) = P(\nabla_u u)$ since P annihilates exact forms. (Remember P projects onto the divergence free part).

So substituting into the Euler equations, we get the following alternative form:

$$\frac{\partial \tilde{u}}{\partial t} + P(L_u \tilde{u}) = 0. \tag{*}$$

Let η_t be the flow of u_t. Then $\ell(t) = \eta_t(\ell)$, and so changing variables,

$$\int_{\eta_t(\ell)} \tilde{u}_t = \int_\ell \eta_t^*(\tilde{u}_t)$$

which becomes, on carrying out the differentiation,

$$\frac{d}{dt}\int_{\eta_t(\ell)} \tilde{u}_t = \int_\ell \eta_t^*(L_u\tilde{u}) + \eta_t^* \frac{\partial \tilde{u}}{\partial t} .$$

Let $P(L_u\tilde{u}) = L_u\tilde{u} - \text{grad } q$. By Stokes theorem $\int_\ell \text{grad } q = 0$,

$$\frac{d}{dt}\int_{\ell(t)} \tilde{u}_t = \int_\ell \eta_t^*(L_u\tilde{u} + \frac{\partial \tilde{u}}{\partial t} - \text{grad } q) = 0 . \quad \square$$

In practical fluid mechanics, this is an important theorem. One can obtain a lot of qualitative information about specific flows by following a closed loop throughout time and using the fact the circulation is constant.

The quantity $d\tilde{u} = \omega$ is the <u>vorticity</u>. (In three dimensions $\vec{\omega} = \vec{\nabla} \times \vec{u}$.) From (*) we get $\frac{\partial \omega}{\partial t} + L_u\omega = 0$ and so $\omega_0 = \eta_t^*\omega_t$, showing that vorticity moves with the fluid. This is, via Stokes theorem, another way of phrasing Kelvin's theorem.

Steady Flows.

A flow is steady if its vector field satisfies $(\partial u/\partial t) = 0$, i.e., u is constant in time. This condition means that the "shape" of the fluid flow is not changing. Even if each particle is moving under the flow, the global configuration of the fluid does not change.

Not much is really known about steady flows, their stability, or what initial conditions result in steady flows. We should mention, however, that for viscous flow quite a bit more is known. See for example Ladyzhenskaya [2] and Finn [1]. There are some elementary equivalent formulations of the Euler problem.

<u>Proposition.</u> <u>Let</u> u_t <u>be a solution to the Euler equations on a manifold</u> M <u>and</u> η_t <u>its flow. Then the following are equivalent:</u>

(1) $u_0 \in T_e \mathcal{D}_\mu^s$ <u>yields a steady flow</u> (i.e., $(\partial u/\partial t) = 0$)

(2) η_t <u>is a one parameter subgroup of</u> $\mathcal{D}_\mu^s(M)$

(3) $L_{u_0} \tilde{u}_0$ <u>is an exact form</u>

(4) $i_{u_0} du_0$ <u>is an exact form.</u>

The details are omitted.

It follows at once from (4) that if $u_0 \in T_e \mathcal{D}_\mu^s(M)$ is a <u>harmonic</u> vector field; i.e., u_0 satisfies $\delta u_0 = 0$ and $\widetilde{du_0} = 0$ then it yields a stationary flow. Also it is known there are other steady flows for manifolds with boundary. For example, on a closed 2-disc, with polar coordinates (r, θ), $v = f(r)(\partial/\partial\theta)$ is the velocity field of a steady flow because

$$\nabla_v v = -\nabla p \ , \quad \text{where} \quad p(r, \theta) = \int_0^r f^2(s) s \, ds \ .$$

Clearly such a v need not be harmonic.

For the remainder of this chapter we shall fill in a number of details. In particular we shall prove $\mathcal{D}^s_\mu$ is a smooth manifold, will prove Arnold's theorem and outline the proof that the geodesic spray is smooth. (In Arnold [1], and Marsden-Abraham [1], the result of Arnold is proved using Lie group methods).

Groups of Diffeomorphisms.

These objects have a very interesting yet complicated structure. For this section we let M be a compact manifold without boundary. Let $\mathcal{D}^s(M) = \{f \in H^s(M, M) | f$ is one-one, orientation preserving and $f^{-1} \in H^s(M, M)\}$. The fact that $\mathcal{D}^s(M)$ is a manifold is a trivial consequence of the fact that $H^s(M, M)$ is a manifold and the following proposition;

<u>Proposition</u>. <u>If</u> $s > (n/2) + 1$, <u>then</u> $\mathcal{D}^s(M)$ <u>is open in</u> $H^s(M, M)$.

<u>Proof</u>. Since $s > (n/2) + 1$, we have a continuous inclusion $H^s(M, M) \subset C^1(M, M)$ (by the Sobolev Theorem). So it is sufficient to show that if a map g on M is C^1 close to a diffeomorphism, then g is a diffeomorphism. To show this, note that $G : f \mapsto \inf_{x \in M} Jf(x)$ is a continuous real valued map on $C^1(M, M)$, where $Jf(x)$ is the Jacobian of $f : f^*\mu = (Jf)\mu$. Also, since M is compact, if $f \in \mathcal{D}^s(M)$, then $G(f) \neq 0$. By continuity of G, there is a neighborhood U of f in $C^1(M, M)$ such that if $g \in U$ then $G(g) \neq 0$. By the inverse function theorem U consists of local diffeomorphisms. It is easy to show that if $g \in U$ then g is an onto map. This is

because g(M) is open in M, as g is a local diffeomorphism and since g is continuous and M is compact, then g(M) is closed. Hence if M is connected g(M) = M. (If M is not connected, one need just remark that g maps into each component of M since f does and g is uniformly close to f.) It remains to show there is a neighborhood of f containing only 1-1 functions. (It is not true that a local diffeomorphism on a compact set is a diffeomorphism. Consider the map which wraps S^1 around itself twice.) It is an easy exercise in point set topology to show that if M is connected then any local diffeomorphism on M is a covering map; that is, is globally k to 1 for some integer k. Also, the function that assigns to a local diffeomorphism f the number of elements in $f^{-1}(x)$ for any $x \in M$ is continuous in the C^1 topology onto the integers. In particular there is a neighborhood of a diffeomorphism containing only diffeomorphisms. □

Because of the above proposition, we will henceforth assume $s > (n/2) + 1$.

It is unknown whether, in general, the composition of two H^s maps is again H^s. In all known proofs one needs that one of the maps is a diffeomorphism or is C^∞. The main composition properties are stated in the following.

<u>Theorem.</u> (a) $\mathcal{D}^s$ <u>is a group under composition.</u>

(b) (α-<u>Lemma</u>) <u>If</u> $\eta \in \mathcal{D}^s$ <u>the map</u> $R_\eta : \mathcal{D}^s \to \mathcal{D}^s$,

$\zeta \mapsto \zeta \circ \eta$ is a C^∞ map (in fact R_η is clearly "formally linear" and continuous).

(c) (ω-Lemma-Global) If $\eta \in \mathcal{D}^s$, then $L_\eta : \zeta \mapsto \eta \circ \zeta$ is C^0. (This map is definitely not smooth, in fact it is not even a locally Lipschitz map.*)

(c)' More generally, the map

$$\mathcal{D}^{s+\ell} \times \mathcal{D}^s \to \mathcal{D}^s$$
$$(\eta, \zeta) \mapsto \eta \circ \zeta$$

is C^ℓ.

(d) $\mathcal{D}^s$ is a topological group.

Remark. (d) follows from the other parts of the theorem because of the following lemma of Montgomery [1]:

Lemma. Let G be a group that is also a topological space. Assume further that G is a separable, metrizable, Baire space and multiplication in G is separately continuous. Then G is a topological group.

We shall not prove (a), (b), (c)', here since we have already given the basic ideas involved. The proof may be found in Ebin [1]. Another useful fact proved by Ebin is that if η is an H^s map with a C^1 inverse, then the inverse is H^s. This is analogous to what one has in the C^k inverse function theorem (Lang [2]). These results also extend to the L_p^k and $C^{k+\alpha}$ spaces; cf. Bourguinon and Brezis [1] and Ebin-Marsden [1].

---

* That it cannot be locally Lipschitz follows from an example given by T. Kato. See the footnote on page 118.

$\mathcal{D}^s$ as a "Lie group".

$\mathcal{D}^s$ is not precisely a Lie group, (since a left multiplication is continuous, but not smooth) but it shares some important Lie group properties. If we were to work with $\mathcal{D} = \mathcal{D}^\infty$, we would have Lie group, but not a Banach manifold.

In general if G is a Lie group and $e \in G$ is the unit element, then the Lie Algebra $\mathcal{G}$ of G may be identified with $T_e G$. Hence, $T_e \mathcal{D}^s(M) = \mathfrak{X}^s(M) = H^s(TM) = H^s$ vector fields on M (recall members of $T_e \mathcal{D}^s(M)$ cover the identity map on M) serve as the Lie algebra for $\mathcal{D}^s$. Since right multiplication is smooth, we can talk about right invariant vector fields on $\mathcal{D}^s$. By the ω-lemma, if $X \in \mathfrak{X}^{s+\ell}$, the map $\widetilde{X} : \eta \mapsto X \circ \eta$ is a C^ℓ map from $\mathcal{D}^s$ to $T\mathcal{D}^s$ ($\ell \geq 0$); in particular $X \circ \eta \in T_\eta \mathcal{D}^s$ and so it is a vector field on $\mathcal{D}^s$. In fact $\widetilde{X}$ is a right invariant C^ℓ vector field on $\mathcal{D}^s$ (i.e., $(R_\eta)_* (\widetilde{X}_\zeta) = \widetilde{X}_\zeta \circ \eta$ for $\eta \in \mathcal{D}^s$ and $\widetilde{X}_\zeta = \widetilde{X}(\zeta) \in T_\zeta \mathcal{D}^s$). Conversely if $\widetilde{X}$ is a right invariant C^ℓ vector field, then $\widetilde{X}(e) \in \mathfrak{X}^{s+\ell}$. In fact the right invariant C^ℓ vector fields are isomorphic to $\mathfrak{X}^{s+\ell}$ by evaluation at e, and in particular $T_e \mathcal{D}^s$ is isomorphic to the C^0 right invariant vector fields.

For $\ell \geq 1$, there is a natural Lie bracket operation on the C^ℓ right invariant vector fields on $\mathcal{D}^s$. This defines the bracket operation on the corresponding members of $T_e \mathcal{D}^s(M)$. We now establish that the Lie algebra structure of $\mathcal{D}^s$ is the usual Lie algebra structure on the vector fields.

Theorem. Let $\ell \geq 1$ and for $X, Y \in H^{s+\ell}(TM)$, let $\widetilde{X}$ and $\widetilde{Y}$ be the corresponding right invariant vector fields on $\mathcal{D}^s$. Then $[\widetilde{X}, \widetilde{Y}]_e = [X, Y]$, the usual Lie bracket of vector fields on M.

Proof. Recall that locally $[X, Y] = DX \cdot Y - DY \cdot X$ (where DX is the derivative of X; cf. lecture 1. However, as shown above, for $\eta \in \mathcal{D}^s$, $\widetilde{X}(\eta) = X \circ \eta$ and $\widetilde{Y}(\eta) = Y \circ \eta$, so in particular since $T\widetilde{Y} \cdot X = TY \cdot X$ we get $[\widetilde{X}, \widetilde{Y}]_e = (D\widetilde{X} \cdot \widetilde{Y} - D\widetilde{Y} \cdot \widetilde{X})_e = D\widetilde{X}(e) \cdot \widetilde{Y}(e) - D\widetilde{Y}(e)\widetilde{X}(e) = DX \cdot Y - DY \cdot X$. □

Note since $DX \cdot Y \in H^{s+\ell-1}(TM)$, we really cannot put this bracket on $T_e \mathcal{D}^s = \mathfrak{X}^s$ and none of the $\mathcal{D}^{s+\ell}(TM)$ are Lie algebras since they are not closed under the bracket operation; one would have to pass to $\mathcal{D} = \mathcal{D}^\infty$.

For any Lie group G, there is a standard exp map from $\mathcal{J}$ onto a neighborhood of the identity e in G. If $X \in \mathcal{J}$, there is a unique one parameter smooth subgroup c in G (i.e., $c(t+s) = c(t) \cdot c(s)$ and $c(0) = e$) such that $c'(0) = X$. In this case X is the infinitesimal generator of c; c is the solution of $c'(t) = \widetilde{X}(c(t))$ where $\widetilde{X}$ is the right invariant vector field equaling X at e. Define $\exp(X) = c(1)$.

If G has a Riemannian structure, then there is another map $\overline{\exp} : \mathcal{J} \to G$ defined (as above) by following geodesics instead of subgroups. If the metric is bi-invariant (i.e., if $g = (g_{ij})$ is the Riemannian metric, then for $a \in G$, $(R_a)^*(g) = (L_a)^*(g) = g$)

then it is easy to show the two exp maps coincide.

In the case of $\mathcal{D}^s$, we will construct a metric that is right invariant, but not left invariant, and so the two exp maps will in general be different.

Actually $\mathcal{D}^s$ (and $\mathcal{D}^s_\mu$) have no bi-invariant metrics. (Indeed, as in Sternberg [1], a group G has a bi-invariant metric iff the image of G under the adjoint map is relatively compact.)

Let $X \in T_e \mathcal{D}^s$. Then $X \in H^s(TM) = \mathfrak{X}^s$ and therefore has a flow F_t ($F_t(m)$ is the integral curve of X starting at m). This is a one parameter group since $F_{s+t} = F_s \circ F_t$. Since M is compact, F_t is defined on all of M for all $t \in R$. (Flows of C^r vector fields on compact manifolds are always complete.)

Let us argue that we should have

$$\exp X = F_1$$

where exp is the (right) exponential map on $\mathcal{D}$. Indeed we need to show F_t is an integral curve of $\widetilde{X}$ defined above. But

$$\frac{d}{dt} F_t(m) = X(F_t(m))$$

i.e.

$$\frac{d}{dt} F_t = X \circ F_t = \widetilde{X}(F_t) .$$

Hence F_t is an integral curve in $\mathcal{D}^s$ of $\widetilde{X}$. This justifies us in saying that $\exp X = F_1$.

Actually it is not obvious that $F_t \in \mathcal{D}^s$; i.e., the flow of an H^s vector field is H^s. (This, of course, is well-known in the C^k case -- see lecture 1.) However the H^s version is also true. See Ebin-Marsden [1], Bourguinon and Brezis [1] and Fischer-Marsden [2] for proofs.

So via this theorem and the remark that $F_0 = \text{id}$, we have a sort of Lie group exponential map from $T_e \mathcal{D}^s(M)$ <u>into</u> a neighborhood of identity, $X \mapsto F_1$. It is natural to ask why not use this exp map to directly define charts on $\mathcal{D}^s(M)$. We cannot do this because it is a fact that exp does <u>not</u> map onto any neighborhood of the identity in $\mathcal{D}^s(M)$. This is equivalent to saying that there are diffeomorphisms near e not embeddable in a flow. In other words for any neighborhood U of e in $\mathcal{D}^s$, there is $\eta \in U$ such that there is no flow F_t with $F_1 = \eta$. In fact η will not, in general, have a square root. Explicit examples have been given by several people such as Eells and Smale. One is written down in Omori [1] and in Friefeld [1].

A consequence of this is that the exp map on $T_e \mathcal{D}^s$ is <u>not</u> C^1, for if it were, it would be locally onto by the inverse function theorem.

<u>Volume Preserving Diffeomorphisms</u>.

For now let M be a compact Riemannian manifold without boundary. (The boundary case is done below.) Let μ be the volume form given by the metric on M. Recall from the introduction that $\mathcal{D}_\mu^s = \{f \in \mathcal{D}^s | f^*(\mu) = \mu\}$. We shall show that $\mathcal{D}_\mu^s$ is a smooth submanifold of $\mathcal{D}^s$.

Recall that if $f : P \to Q$ is a smooth map between manifolds, f is a <u>submersion</u> on a set $A \subset P$ if $T_x f : T_x P \to T_{f(x)} Q$ is a surjection, for each $x \in A$ and the kernel splits. We showed in lecture one that if P, Q are Hilbert manifolds and $f : P \to Q$ is a C^∞ map, then for $g \in Q$, $f^{-1}(g)$ is a C^∞ submanifold of P, if f is a submersion on $f^{-1}(g)$.

We shall need the following:

<u>Lemma</u>. <u>Let λ be an n-form on M such that $\int_M \lambda = 0$. Then λ is exact; $\lambda = d\alpha$ for an n-1 form α.</u>

This is a special case of de Rham's theorem, stating that a closed form is exact if all its periods vanish. For the proof, see for example Warner [1]. A discussion is also found in Flanders [1].

<u>Theorem</u>. <u>Let $s > (n/2) + 1$. Then $\mathcal{D}_\mu^s$ is a closed C^∞ submanifold of $\mathcal{D}^s$.</u>

<u>Proof</u>. Let μ be the volume form on M. By the Hodge theorem, $[\mu]_s = \mu + d(H^{s+1}(\Lambda^{n-1}))$ is a closed affine subspace of $H^s(\Lambda^n)$, being the translate of the closed subspace $d(H^{s+1}(\Lambda^{n-1}))$ by μ. Define the map

$$\psi : \mathcal{D}^{s+1}(M) \to [\mu]_s$$

$$\eta \mapsto \eta^*\mu .$$

Now $\eta^*\mu \in [\mu]_s$ since

$$\int_M (\mu - \eta^*\mu) = \int_M \mu - \int_M \eta^*\mu = 0 .$$

Hence $\mu - \eta^*\mu = d\alpha$ by the lemma. By the ω-Lemma, one can easily see that ψ is a C^∞ map. Now $\mathcal{D}^{s+1}_\mu(M) = \psi^{-1}(\mu)$, so if ψ is a submersion then $\mathcal{D}^{s+1}_\mu(M)$ is a C^∞ submanifold of $\mathcal{D}^{s+1}(M)$.

We shall show this at $e \in \mathcal{D}^s(M)$ (e is the identity map). It turns out that $T_e\psi(X) = L_X\mu$ where $X \in T_e\mathcal{D}^{s+1}(M)$. Indeed let η_t be a curve tangent to X, such as its flow. Then $T_e\psi(X) = (d/dt)\eta_t^*\mu|_{t=0}$ which is indeed the Lie derivative. Using the "magic" formula $L_X\mu = di_X\mu + i_Xd\mu$ for the Lie derivative and the fact that $d\mu = 0$, we get

$$T_e\psi(X) = L_X\mu = di_X\mu .$$

Hence to show $T_e\psi$ is a surjection, we only need show that

$$\{i_X\mu : X \in T_e\mathcal{D}^{s+1}\} = H^{s+1}(\Lambda^{n-1}) .$$

But $i_X\mu = *\widetilde{X}$ and $*$ is a bijection between $n-1$ forms and 1-forms. Hence $T_e\psi$ is onto. Similarly $T_\eta\psi$ is onto. □

However this last step in the proof only holds if μ is a (nowhere zero) n-form or a <u>closed</u> nondegenerate 2-form. This remark allows us to show that the diffeomorphisms that preserve a symplectic form form a submanifold of the diffeomorphism group using the same sort of argument.

It follows from the basic connection between Lie derivatives and flows given in Lecture one that a vector field generates volume preserving diffeomorphisms if and only if it is divergent free. In our context this means $T_e \mathcal{D}^s_\mu(M) = \{X \in \mathcal{D}^s(M) \mid \delta\widetilde{X} = 0\}$. This is clearly a subspace of $T_e \mathcal{D}^s(M)$ and in fact $T_e \mathcal{D}^s_\mu(M)$ is closed under the bracket operation, in the same sense as $T_e \mathcal{D}^s(M)$ (see page 92 above).

Manifolds with Boundary.

Suppose M is a compact, oriented, Riemannian Manifold with smooth boundary. Let $\widetilde{M}$ be the double of M, i.e., $\widetilde{M}$ is two copies of M with the boundaries identified, with the obvious differential structure. Now $\widetilde{M}$ is a compact, oriented, Riemannian manifold without boundary and M has a natural imbedding in $\widetilde{M}$. We have the manifold structure of $H^s(M, \widetilde{M})$ by our above work. Clearly $\mathcal{D}^s(M) \subset H^s(M, \widetilde{M})$ and in fact:

<u>Theorem.</u> $\mathcal{D}^s(M)$ <u>is a</u> C^∞ <u>submanifold of</u> $H^s(M, \widetilde{M})$.

<u>Sketch of Proof.</u> Briefly, we put a metric on $\widetilde{M}$ such that $\partial M \subset \widetilde{M}$ is totally geodesic. Then let $E : TH^s(M, \widetilde{M}) \to H^s(M, \widetilde{M})$ be the exponential map associated with this metric.

Let $\eta \in \mathcal{D}^s(M) \subset H^s(M, \widetilde{M})$ and choose an exponential chart

$E : U \subset T_\eta H^s(M, \widetilde{M}) \to H^s(M, \widetilde{M})$ about η. Also we should have

$$T_\eta \mathcal{D}^s(M) = \{X \in H^s(M, \widetilde{TM}) \mid X \text{ covers } \eta \text{ and}$$

$$X(x) \in T_{\eta(x)} \partial M \text{ for all } x \in \partial M\}$$

which is a closed subspace of

$$H^s(M, \widetilde{TM}) = TH^s(M, \widetilde{M}) .$$

Since ∂M is totally geodesic, E takes $U \cap T_\eta \mathcal{D}^s(M)$ onto a neighborhood of η in $\mathcal{D}^s(M)$. See Ebin-Marsden [1] for details. □

By inspecting the above argument we see $T_e \mathcal{D}^s(M) = \{H^s$ vector fields on M that are tangent to $\partial M\}$. Formally, this is a Lie algebra in the same sense as we had when M had no boundary.

<u>Theorem.</u> <u>If μ is the volume on M and $\mathcal{D}^s_\mu(M)$ is the set of volume preserving diffeomorphisms, then $\mathcal{D}^s_\mu(M) \subset \mathcal{D}^s(M)$ is a smooth submanifold.</u>

This is proven as in the case that M has no boundary. This proof works here because we have the Hodge theorems for manifolds with boundary. The rest of the material from the no boundary case (such as the α and ω-lemmas) carries over to the case when M has a boundary. For the non-compact case, see Cantor [1,2].

If M has boundary, then $H^s(M,M)$ will not be a smooth manifold, but will have "corners". Thus it is interesting that nevertheless, $\mathcal{D}^s(M)$ is a smooth manifold.

Topology of the Diffeomorphism Group.

For topological theorems we can work in $\mathcal{D}(M) = \mathcal{D}^\infty(M)$. Indeed it follows from very general results of Cerf [1] and Palais [3] that the topology of $\mathcal{D}^s$ and $\mathcal{D}$ are the same; one uses the fact that the injection of $\mathcal{D}$ into $\mathcal{D}^s$ is dense. The first theorem in this field was proven by Smale [1] in 1959. He showed that $\mathcal{D}(S^2)$ is contractable to $SO(3)$; here S^2 is the 2-sphere, and $SO(3)$ is the special orthogonal group on R^3, which we can regard as the (identity component of the) isometry group of S^2. This theorem was extended to all compact 2-manifolds by Earle and Eells [1] and to the boundary case by Earle and Schatz [1].

It is fairly simple to show that $\mathcal{D}(S^1)$ <u>is contractable to</u> $SO(2)$. The following argument is based on a suggestion of J. Eells.

First fix $s \in S^1$. Let $\theta : [0, 1] \to S^1$ be a parameterization of S^1 such that $\theta(0) = \theta(1) = s$. Now let f be a diffeomorphism that leaves s fixed. Then the map

$$h_f(t, x) : [0, 1] \times S^1 \to S^1$$

$$(t, x) \mapsto \begin{cases} \theta(t\theta^{-1}(x) + (1-t)\theta^{-1}(f(x))) & x \neq s \\ s & x = s \end{cases}$$

is an homotopy from f to id_{S^1}.

Suppose $g : S^1 \to S^1$ maps s to $g(s) \neq s$; then there is a rotation $r : S^1 \to S^1$ that carries $g(s)$ to s and therefore

$r \circ g(s) = s$. Hence, by the above argument $r \circ g$ is homotopic to the identity. Therefore g is homotopic to r^{-1}, which is, naturally, also a rotation. □

For dimension 3 the situation is much more complicated and little is known. The work of Cerf [2] seems indicative of the complexity. Antoneli et al. [1] have shown that if M has high dimension $\mathcal{D}(M)$ will not have the homotopy type of a finite cell complex. Various people have also been working towards showing $\mathcal{D}(M)$ is a simple group; cf. Herman [1], Epstein [1] and Herman-Sergeraert [1]. This result was actually known to von Neumann for the case of homeomorphisms. It has recently been announced for $\mathcal{D}(M)$ by W. Thurston.

Another important result in this field is that of Omori [1]. He proved that for any compact Riemannian manifold without boundary $\mathcal{D}(M)$ is contractable to $\mathcal{D}_\mu(M)$, the set of volume preserving diffeomorphisms. In fact if $V = \{\nu \in C^\infty(\Lambda^n) \mid \nu$ is nondegenerate, positively oriented and $\int_M \nu = \int_M \mu\}$ ($C^\infty(\Lambda^n)$ are the C^∞ n-forms) then $\mathcal{D}(M)$ is diffeomorphic to $\mathcal{D}_\mu \times V$. This implies $\mathcal{D}(M)$ is contractable to $\mathcal{D}_\mu(M)$ since V is contractable to μ. (In fact V is convex.) The proof that $\mathcal{D}(M) \approx \mathcal{D}_\mu(M) \times V$ uses an important result of Moser [1].

Theorem. [Moser]. *If on a compact manifold* M *, there are* 2 *volume elements* μ *and* ν *such that* $\int_M \nu = \int_M \mu$ *, then there is map* $f \in \mathcal{D}(M)$ *such that* $f^*(\nu) = \mu$.

We formulate the results following Ebin-Marsden [1].

<u>Theorem.</u> Let M <u>be compact without boundary with a smooth volume element</u> μ . <u>Let</u>

$$V = \{\nu \in C^\infty(\Lambda^n) \mid \nu > 0, \int_M \nu = \int_M \mu\} .$$

<u>Then</u> $\mathcal{D}$ <u>is diffeomorphic to</u> $\mathcal{D}_\mu \times V$. <u>In particular (since</u> V <u>is convex),</u> $\mathcal{D}_\mu$ <u>is a deformation retract of</u> $\mathcal{D}$.

For the proof, we begin by proving Moser's result.

<u>Lemma.</u> <u>There is a map</u> $\chi : V^s \to \mathcal{D}^s$, $s > (n/2) + 1$ <u>such that</u> $\psi_\mu : \mathcal{D}^s \to V^{s-1}$, $\psi_\mu(\eta) = \eta^*(\mu)$ <u>satisfies</u> $\psi_\mu \circ \chi =$ <u>identity. Further,</u> $\chi : V \to \mathcal{D}$ <u>is a</u> C^∞ <u>map.</u>

<u>Proof.</u> For $\nu \in V_s$, let $\nu_t = t\nu + (1-t)\mu$, so that $\nu_t \in V^s$. Since $\int \mu = \int \nu$, we can write, as before, $\mu - \nu = d\alpha$. Define X_t by $i_{X_t}\nu_t = \alpha$ so that $X_t \in H^s(TM)$. Let η_t be the flow of X_t, so $\eta_t \in \mathcal{D}^s$. Define $\chi(\nu) = \eta_1^{-1}$. We want to show that $\eta_t^*(\nu_t) = \mu$ by showing $d/dt(\eta_t^*(\nu_t)) = 0$. Indeed, we have, from the basic fact about Lie derivatives

$$\frac{d}{dt}(\eta_t^*(\nu_t)) = \eta_t^*(L_{X_t}\nu_t + \frac{d}{dt}\nu_t)$$

$$= \eta_t^*(di_{X_t}\nu_t - (\mu - \nu))$$

$$= \eta_t^*(d\alpha - \mu + \nu) = 0 . \quad \square$$

Note that χ is canonically defined, given the Riemannian metric on M .

Proof. Define $\Phi : \mathcal{D}_\mu \times V \to \mathcal{D}$ by $\Phi(\zeta, \nu) = \zeta \circ \chi(\nu)$. Then $\Phi^{-1}(\eta) = (\eta \circ (\chi \eta^*)(\mu))^{-1}, \eta^*(\mu))$ as is easily checked. □

This can be generalized to the boundary case as well.

The basic technique used here is essentially the same as that used in the proof of Darboux's theorem in lecture 2.

It is also possible to study other groups of diffeomorphisms. For example, let M be a compact manifold and let G be a compact group. Let $\Phi : G \times M \to M$ be a group action, and let $\Phi_g(m) = \Phi(g, m)$. Set

$$\mathcal{D}_\Phi^s(M) = \{\eta \in \mathcal{D}^s(M) | \eta \circ \Phi_g = \Phi_g \circ \eta\} .$$

This is a subgroup of $\mathcal{D}^s(M)$, a C^∞ submanifold and has "Lie algebra"

$$T_e \mathcal{D}_\Phi^s(M) = \{V \in T_e \mathcal{D}^s(M) | V \text{ commutes with all}$$
$$\text{infinitesimal generators of } \Phi\} .$$

Of course, we can also take $\mathcal{D}_{\mu,\Phi}^s(M) = \mathcal{D}_\mu^s(M) \cap \mathcal{D}_\Phi^s(M)$. Since this intersection is not in general transversal, it is not obvious that $\mathcal{D}_{\mu,\Phi}(M)$ is a submanifold. It is true, but requires some argument (Marsden [7]). The group $\mathcal{D}_{\mu,\Phi}^s(M)$ is important in the study of flows with various symmetries (e.g., a flow in R^3 that is symmetric with respect to a given axis). Also, in general we find that $\dim(\mathcal{D}_\Phi(M))$ and $\text{codim}(\mathcal{D}_\Phi(M))$ are both infinite so Frobenius methods do not work

(Leslie [2] and Omori [1, 3] have shown that if $\mathcal{J}$ is a Lie subalgebra of $T_e \mathcal{D}$ with finite dimension or codimension, then $\mathcal{J}$ comes from a smooth subgroup of $\mathcal{D}$).

The metric on $\mathcal{D}_\mu^s$.

It follows from the results we established above that the tangent space to $\mathcal{D}_\mu^s(M)$ at a point $\eta \in \mathcal{D}_\mu^s$ is given by
$T_\eta \mathcal{D}_\mu^s(M) = \{X \in H^s(M, TM) \mid X \text{ covers } \eta,\ \delta(X \circ \eta^{-1}) = 0,\ \text{and } X \text{ is parallel to } \partial M\}$. Note that if $X \in T_\eta \mathcal{D}_\mu^s(M)$ then $X \circ \eta^{-1}$ is a vector field on M. If we are working on $\mathcal{D}^s$ then the divergence condition $\delta(X \circ \eta^{-1}) = 0$ is dropped, so $T_\eta \mathcal{D}^s$ consists of H^s sections parallel to ∂M which cover η.

Let M be a compact Riemannian manifold $m \in M$ and let $\langle\ ,\ \rangle_m$ be the inner product on $T_m M$. Now we put a metric on $\mathcal{D}^s(M)$ as follows: Let $\eta \in \mathcal{D}^s(M)$ and $X, Y \in T_\eta \mathcal{D}^s(M)$. Then $X(m)$ and $Y(m)$ are in $T_{\eta(m)} M$. Now define:

$$(X, Y)_\eta = \int_M \langle X(m), Y(m) \rangle_{\eta(m)}\, d\mu(m).$$

This is a symmetric bilinear form on each tangent space $T_\eta \mathcal{D}^s$ of $\mathcal{D}^s(M)$. By restriction it also defines a symmetric bilinear form on each tangent space of $\mathcal{D}_\mu^s$.

The norm induced by this inner product is clearly an L^2 norm and hence the topology it induces is weaker than the H^s topology on each $T_\eta \mathcal{D}^s(M)$. Thus, in the terminology of lecture 2, $(\ ,\ )$ is a weak metric. It is important to allow weak metrics although most definitions of Riemannian manifolds exclude this (as in Lang [1]). Also, recall that this is the physically appropriate metric for hydrodynamics, since for $X \in T_\eta \mathcal{D}^s(M)$, $\frac{1}{2}(X, X)$ represents the total kinetic energy of a fluid in state η and velocity field $v = X \circ \eta^{-1}$. So finding geodesics is formally the same as finding a flow satisfying a least energy condition. (This is the connection with variational principles or least action principles in fluid mechanics.)

This metric $(\ ,\ )_\eta$ just constructed is smooth in this sense: If $B(T_\eta \mathcal{D}^s_\mu, T_\eta \mathcal{D}^s_\mu)$ is the vector bundle of bilinear maps over the tangent spaces of $\mathcal{D}^s(M)$ (i.e., if $g_\eta \in B(T_\eta \mathcal{D}^s_\mu, T_\eta \mathcal{D}^s_\mu)$ then $g_\eta : T_\eta \mathcal{D}^s_\mu(M) \times T_\eta \mathcal{D}^s_\mu(M) \to \mathbb{R}$ is bilinear), then the map $\eta \mapsto (\ ,\ )_\eta$ is a section of this bundle, and to say the metric is smooth is to say this section is smooth. (Here each fiber of $B(T_\eta \mathcal{D}^s_\mu, T_\eta \mathcal{D}^s_\mu)$ has the standard topology put on bilinear maps on banach spaces, and one constructs the bundle as in Lang [1], Ch. III, §4.)

Note. It is not always true that a weak metric yields geodesics. For example, suppose $\partial M \neq \emptyset$. Then on $\mathcal{D}^s(M)$, this weak metric would yield geodesics which would try to cross the boundary of M. We shall see this in more detail below.

The Spray on $\mathcal{D}^s_\mu$.

We now wish to construct the spray on $\mathcal{D}^s_\mu$ corresponding to the metric $(\ ,\ )$. Recall from lecture 2 that this means finding the Hamiltonian vector field on $T\mathcal{D}^s_\mu$ corresponding to the energy

$K(X) = \frac{1}{2}(X, X)$. Assume

Theorem. *Let* Z *be the spray of the metric on* M. *Then the spray of* $(\,,\,)$ *on* $\mathcal{D}^s(M)$ *is given by*

$$\bar{Z} : T\mathcal{D}^s \to T^2(\mathcal{D}^s) \,;\, X \mapsto Z \circ X \,.$$

We shall just make the result plausible, leaving details to the reader. See also Ebin-Marsden [1] and Eliasson [1].

Note. As with $T\mathcal{D}^s$, it is not hard to see that $T_X(T\mathcal{D}^s)$ consists of H^s maps $Y : M \to T^2 M$ which cover X; i.e., such that $\pi_1 \circ Y = X$, where $\pi_1 : T^2 M \to TM$ is the projection. The spray Z satisfies $\pi_1 \circ Z =$ identity, since Z is a vector field. Thus $Z(X) \in T_X T\mathcal{D}^s$ so $\bar{Z}$ is indeed a vector field on $T\mathcal{D}^s$.

The idea behind the proof is to realize that we can explicitly write down what should be the geodesics on $\mathcal{D}^s(M)$. From the construction of charts on $\mathcal{D}^s(M)$, there is the map $\overline{\exp} : T\mathcal{D}^s(M) \to \mathcal{D}^s(M)$ where $\overline{\exp}(X) = \exp \circ X$ and $\exp : TM \to M$ is the Riemannian exponential map on M. First we assert that for $X \in T_e \mathcal{D}^s(M)$, the geodesic on $\mathcal{D}^s(M)$ through e in the direction X is given by $t \mapsto \overline{\exp}(tX)$. What this geodesic looks like is seen by considering any $m \in M$. Then $t \mapsto \overline{\exp}(tX)(m) = \exp(tX_m)$ is the geodesic starting at m in the direction X_m. So $\overline{\exp}(tX)$ represents all of the geodesics on M in the direction of the vector field X evaluated at $m \in M$. Now in general, as t increases it is likely that some pair of geodesics will intersect. Say this happens at $t = t_0$. Then $\overline{\exp}(t_0 X)$ is not a diffeomorphism. Hence even if M is a simple manifold (like the flat 2-torus), $\mathcal{D}^s(M)$ is <u>not</u> geodesicly complete.

If we can show $t \mapsto \exp tX$ is a geodesic on $\mathcal{D}^s$, then the formula for Z follows at once, since for each $m \in M$, $v(t) = (d/dt)\exp(tX(m))$ satisfies $(d/dt)v(t) = Z(v(t))$, and $v(0) = X(m)$. Hence it suffices to establish our assertion concerning the geodesics on $\mathcal{D}^s$.

Of course a fundamental property of geodesics is that they locally minimize length. Suppose we have a family of geodesic curves $t \mapsto \eta(t)(m)$, starting at $m \in M$, where for $t_0 \in R$, t_0 near 0, the map $m \mapsto \eta(t_0)(m)$ is a diffeomorphism so that $t \mapsto \eta_t$ is a curve in $\mathcal{D}^s$. Then since the length of a curve in $\mathcal{D}^s(M)$ given by our weak metric is the integral over M of the lengths of each curve, $t \mapsto \eta_t(m)$, this integrated length is also minimized. Hence it is reasonable that $t \mapsto \eta(t)$ should be a geodesic on $\mathcal{D}^s(M)$. The curves $t \mapsto \overline{\exp(tX)}(m)$ have all the above properties so $t \mapsto \exp(tX)$ should be a geodesic on $\mathcal{D}^s(M)$. This concludes our justification

<u>Corollary</u>. $\bar{Z}$ <u>is a</u> C^∞ <u>vector field on</u> $T\mathcal{D}^s$.

This is a consequence of the omega lemma since Z is a C^∞ map.

Let us consider a simple example. Let $\mathcal{T}^2$ be the flat 2-torus. Then $T(\mathcal{T}^2) \cong \mathcal{T}^2 \times R^2$ is also a flat 4-manifold and $T(T\mathcal{T}^2) \cong (\mathcal{T}^2 \times R^2) \times (R^2 \times R^2)$. In this case the spray for the flat metric is given by

$$Z : T(\mathcal{T}^2) \to T(T(\mathcal{T}^2)) : (x, v) \mapsto ((x, v), (v, 0)) .$$

The x in the first coordinate is just the base point of the tangent vector in $T\mathcal{T}^2$. The v in the third coordinate is an important formal property of sprays reflecting the fact that the geodesic equations are "second order" (see Lang [1]) and the 0 in the last coordinate reflects the fact that the metric is flat, hence each $\Gamma^i_{jk} = 0$. In this case the geodesics are of the form $\eta(t)(m) = m + tX(m)$ (where $X \in T_e \mathcal{D}^s(\mathcal{T}^2)$ and using the obvious identification). These are straight lines and hence $\mathcal{D}^s(\mathcal{T}^2)$ is "flat". In general, in coordinates $x = (x^1, \ldots, x^n)$ on a manifold M, we have

$$Z(x, v) = ((x, v), (v, -\Gamma^i_{jk} v^j v^k)) .$$

We now consider the metric for $\mathcal{D}^s_\mu(M) \subset \mathcal{D}^s(M)$. Even if $\mathcal{D}^s(M)$ is geometrically relatively simple, as above for $\mathcal{T}^2$, $\mathcal{D}^s_\mu(M)$ may be geometrically very complicated. Consider the above example. It should be clear that the diffeomorphism specified by having each point moving along straight lines is generally not volume preserving. So requiring each point on a geodesic in $\mathcal{D}^s$ to be volume preserving must introduce some curvature. In fact the curvature of the space $\mathcal{D}^s_\mu$ is rather complicated. For $M = \mathcal{T}^2$ it is worked out in Arnold [1].

Suppose S is a submanifold of a Riemannian manifold Q such that we have an orthogonal projection of $T_p Q$ onto $T_p S$ for each $p \in S$. This gives us a bundle map $P : TQ \upharpoonright S \to TS$ (where $TQ \upharpoonright S = \{v_p \in T_p Q | p \in S\}$). This is of course the situation we have for $\mathcal{D}_\mu^s(M)$ as a submanifold of $\mathcal{D}^s(M)$ where the projection is given by the Hodge theorem (i.e., we project onto the divergent free part of X for $X \in T_e \mathcal{D}^s(M)$). In this situation, the following tells us how to put the spray on the submanifold.

<u>Lemma</u>. <u>If Z is the spray on Q then TP∘Z is the spray on S</u>.

This is a standard result in Riemannian geometry, see e.g. Hermann [1]. A proof using Hamiltonian theory may be found in Ebin-Marsden [1].

Now Z is a vector field on TQ as is TP∘Z on TS. However their difference, say h, can be identified (technically by means of the vertical lift -- see below) with a map of TS into TQ ↾ S, which turns out to be (the quadratic part of) the second fundamental form of S as a submanifold. Specifically for $v \in TS$, h(v) is the normal component of $\nabla_v v$; see Hermann [1] or Chernoff-Marsden [1] for details. Thus this difference h in the sprays tells us how curved S is in Q. (More exactly the curvatures on Q and on S are related through this second fundamental form by the Gauss-Codazzi equations; cf. Yano [1], p. 94 and lecture 9.)

Define

$$P_e : T_e \mathcal{D}^s(M) \to T_e \mathcal{D}_\mu^s(M)$$

by carrying a vector field to its divergent free part. As we mentioned above, this is an L^2 orthogonal projection as it is orthogonal for the weak inner product on $T_e \mathcal{D}^s(M)$. We define for $X \in T_\eta \mathcal{D}^s(M)$; $\eta \in \mathcal{D}_\mu^s(M)$

$$P_\eta(X) = (P_e(X \circ \eta^{-1})) \circ \eta .$$

This makes P right invariant and is correct since the metric on $\mathcal{D}_\mu^s(M)$ is right invariant as we now show

Proposition.

(i) Let $\eta \in \mathcal{D}^s(M)$; then $(R_\eta)_* X_\zeta = X_\zeta \circ \eta$ (where $\zeta \in \mathcal{D}^s(M)$, $(R_\eta)_* : T_\zeta \mathcal{D}^s(M) \to T_{\zeta \circ \eta} \mathcal{D}^s(M)$).

(ii) If $\eta \in \mathcal{D}_\mu^s(M)$ then $((R_\eta)_* X, (R_\eta)_* Y)_{\zeta \circ \eta} = (X, Y)_\zeta$, where $X, Y \in T_\zeta \mathcal{D}^s(M)$.

Proof. Part (i) has been used before and is easily seen. We will show the second part. Let $\eta \in \mathcal{D}_\mu^s(M)$; then:

$$((R_\eta)_* X, (R_\eta)_* Y)_{\zeta \circ \eta} = (X \circ \eta, Y \circ \eta)_{\zeta \circ \eta}$$

$$= \int_M <X \circ \eta(m), Y \circ \eta(m)>_{\zeta \circ \eta(m)} d\mu$$

$$= \int_{\eta^{-1}(M)} <X(m), Y(m)>_{\zeta(m)} (\eta^{-1})_*(d\mu) .$$

But, since η^{-1} is volume preserving, $(\eta^{-1})_*(d\mu) = d\mu$ and $\eta^{-1}(M) = M$. Hence

$$((R_\eta)_* X, (R_\eta)_* Y) = \int_M <X(m), Y(m)>_{\zeta(m)} d\mu$$

$$= (X, Y)_\zeta . \quad \square$$

Note that the metric on $\mathcal{D}^s$ is __not__ right invariant.

Putting all this together we can write down the spray S on $\mathcal{D}_\mu^s(M)$. Namely, for $X \in T\mathcal{D}_\mu^s(M)$ we have $S(X) = TP(Z(X)) = TP(Z \circ X)$. There is a major assumption in writing down this formula. When we write TP, we assume P is a C^∞ map. This is not at all obvious since if $X \in T_\eta \mathcal{D}_\mu^s(M)$, we compose X with η^{-1}, project, and then compose with η. As we have seen, composition of H^s maps is not smooth but is at most continuous. However, we have

__Theorem.__ __P is a C^∞ bundle map. That is $P : T\mathcal{D}^s(M) \upharpoonright \mathcal{D}_\mu^s(M) \to T\mathcal{D}_\mu^s(M)$ is C^∞. Hence the spray S on $\mathcal{D}_\mu^s$, $S(X) = TP(Z \circ X)$, is also a C^∞ vector field on $T\mathcal{D}_\mu^s$.__

For a proof see Ebin-Marsden [1]. There is an alternative and perhaps simpler proof to the one in the aforementioned paper. In this proof one defines another metric on $T\mathcal{D}^s(M)$; namely for $X, Y \in T_e \mathcal{D}^s(M)$, set

$$(X, Y)_s = (X, Y) + (\Delta^{s/2} X, \Delta^{s/2} Y)$$

where $(\,,\,)$ is the L^2 metric on $T_e \mathcal{D}^s(M)$, and Δ is the Laplacian. Then extend $(\,,\,)_s$ to make it right invariant.

It turns out that this metric is smooth and by regularity properties of Δ is equivalent to the H^s metric. Smoothness facts like this again are not obvious but are proven in Ebin [1]. These facts are also useful for other purposes. The Hodge decomposition is then easily seen to be orthogonal in this __strong__ metric $(\ ,\ )_s$ and hence it follows automatically that the projection P is smooth.

This result is important for we are going to apply the Picard theorem from ordinary differential equations to the equation:

$$\frac{dX_t}{dt} = S(X_t) = TP(Z \circ X_t)$$

and this requires that S is at least a Lipshitz map.

In case M has boundary, we do not get a spray on $\mathcal{D}^s$, but we do get one on $\mathcal{D}^s_\mu$. This is basically because P projects from vector fields sticking out of M, onto vector fields parallel to ∂M. We shall just accept as plausible that this extension can be made.

As mentioned earlier, it is unknown whether $\mathcal{D}^s_\mu(M)$ is geodesically complete. (By Arnold's theorem, this is the same thing as saying solutions to the Euler equations go for all time, and remain in H^s). Note that this is not equivalent to saying the induced distance metric is complete since the metric is only weak. In fact $\mathcal{D}^s_\mu(M)$ is not complete in this distance sense since the completion of $\mathcal{D}^s_\mu(M)$ under an L^2 topology is much larger than $\mathcal{D}^s_\mu(M)$. (Presumably it consists of a class of measure preserving maps from M to M.)

Derivation of the Euler Equations.

To show geodesics in $\mathcal{D}^s_\mu(M)$ satisfy the Euler equations, we need to know a bit more about T^2M. Let $\pi : TM \to M$ be the projection so that $T\pi : T^2M \to TM$. An element $w \in T^2M$ is called <u>vertical</u> if $T\pi(w) = 0$ (in coordinates this means the third component is 0). Now let $v, w \in T_mM$; define the <u>vertical left</u> of w with respect to v to be

$$(w)_v^\ell = \frac{d}{dt}(v + tw)\Big|_{t=0} \in T_v^2M = T_v(TM) .$$

In coordinates this is simply

$$(w)_v^\ell = (m, v, 0, w) .$$

The proof that geodesics in $\mathcal{D}_\mu^s$ yield solutions to the Euler equations essentially is calculations. The idea is to show that if a curve $X_t \in T\mathcal{D}_\mu^s$ satisfies the spray equation

$$\frac{dX_t}{dt} = S(X_t) , \quad X_t \in T\mathcal{D}_\mu^s(M) ,$$

then X_t gives rise to a solution to the Euler equations in a sense explained below. For alternative proofs, see Arnold [1], Marsden-Abraham [1], or Chernoff-Marsden [1]; see also Hermann [1].

<u>Lemma</u>. $Z(X) = Z \circ X = TX \circ X - (\nabla_X X)_X^\ell$ <u>for</u> $X \in T_e \mathcal{D}^s$.

<u>Proof</u>. In coordinates

$$(\nabla_X X)^i = \sum_j X^j \frac{\partial X^i}{\partial x^j} + \sum_{j,k} \Gamma^i_{jk} X^j X^k .$$

Now

$$(TX \circ X)^i = \sum_j X^j \frac{\partial X^i}{\partial x^j} \quad \text{so} \quad (TX \circ X - (\nabla_X X))^i = -\sum_{jk} \Gamma^i_{jk} X^j X^k .$$

This then puts the right expressions in the fourth component. □

Note that both $TX \circ X$ and $(\nabla_X X)^\ell_X$ are elements of $T_X \mathcal{D}^s_\mu$. The latter is by construction of the vertical lift. To see this for $TX \circ X$, let $\pi_1 : T^2 M \to TM$ be the projection; then since $\pi_1 \circ TX = X \circ \pi$ we have

$$\pi_1 \circ TX \circ X = X \circ \pi \circ X = X$$

since $\pi \circ X$ is the identity.

As we have observed, the map $X \mapsto Z \circ X$ (for $X \in T\mathcal{D}^s(M)$) is C^∞. Hence even though $TX \circ X$ and $\nabla_X X$ are only H^{s-1}, their difference must be H^s.

<u>Lemma</u>. <u>Let σ and X be in $T_\eta \mathcal{D}^s_\mu(M)$ then $TP[(\sigma)^\ell_X] = (P(\sigma))^\ell_X$</u>.

Proof. Since P is linear on each fiber and $P(X) = X$, we get

$$(P(\sigma))^\ell_{P(X)} = \frac{d}{dt}(P(X) + tP(\sigma))\big|_{t=0}$$

$$= \frac{d}{dt} P(X + t\sigma)\big|_{t=0}$$

$$= TP(\frac{d}{dt}(X + t\sigma))\big|_{t=0} \quad \text{(chain rule)}$$

$$= TP[(\sigma)^\ell_X] . \quad \square$$

Lemma. Let $\eta \in \mathcal{D}_\mu^s$ and $X \in T_\eta \mathcal{D}^s(M)$; **then** $TP(T(X \circ \eta^{-1}) \circ X) = \{T(P_e[X \circ \eta^{-1}])\} \circ X$.

Proof. $X \circ \eta^{-1}$ is an H^s vector field on M. Let F_t be its flow (or any curve tangent to X). Let $G_t = (X \circ \eta^{-1}) \circ F_t$. Then $G_0 = X \circ \eta^{-1}$ and $(dG_t/dt) = T(X \circ \eta^{-1}) \circ (X \circ \eta^{-1})$. Thus we get

$$TP(T(X \circ \eta^{-1}) \circ (X \circ \eta^{-1})) = \frac{d}{dt}P(G_t)|_{t=0} \quad \text{(chain rule)}$$

$$= \frac{d}{dt}(P_e(X \circ \eta^{-1}) \circ F_t)|_{t=0}$$

$$= T(P_e(X \circ \eta^{-1})) \circ (X \circ \eta^{-1}).$$

But by right invariance $TP(T(X \circ \eta^{-1}) \circ (X \circ \eta^{-1})) = TP(T(X \circ \eta^{-1}) \circ X) \circ \eta^{-1}$. □

Proposition. *The spray on* $T\mathcal{D}_\mu^s$ *is given by*

$$S(X) = T(X \circ \eta^{-1}) \circ X - (P_e \nabla_{X \circ \eta^{-1}} X \circ \eta^{-1})_X^\ell \circ \eta \quad \text{where } X \in T_{\eta\mu}\mathcal{D}^s(M).$$

Proof. This follows directly from the above lemmas. □

So now that we have an explicit formula for the spray, let us inspect the Euler equations. Recall that these describe the time evolution of the velocity vector field on M. The equations are written out in Eulerian coordinates and are equations involving elements of $T_e \mathcal{D}^s(M)$. The spray on the other hand is a map on all of $T\mathcal{D}^s(M)$. The integral curves of the spray are the velocities written in Lagrangian coordinates. So if $X_t \in T_{\eta(t)}\mathcal{D}^s(M)$ is an integral curve of the spray, we wish to show that the pullback of X_t, i.e., $X_t \circ \eta_t^{-1} \in T_e \mathcal{D}^s(M)$, is a solution of the Euler equations. Let us recall that the vector field $v(t) = X_t \circ \eta_t^{-1}$ is justified as follows. We want η_t to be the flow of v, so this means that

$$\frac{d}{dt}\eta_t(m) = v_t(\eta_t(m)) .$$

Since we are dealing with geodesics and hence $(d\eta/dt) = X$, we get the desired relation $v_t = X_t \circ \eta_t^{-1}$.

It turns out, as we shall see momentarily, that the derivative loss of the Euler equations occurs in this pullback operation (or "coordinate change").

We are interested in computing (dv/dt), and so we need this lemma.

<u>Lemma</u>. We have:

$$\frac{dv(t)}{dt} = \frac{d}{dt}(X_t \circ \eta_t^{-1}) = \frac{dX_t}{dt} \circ \eta_t^{-1} - TX_t \circ T\eta_t^{-1} \circ X_t \circ \eta_t^{-1} .$$

<u>Proof</u>. This follows by differentiating both places t occurs, using the chain rule and the formula

$$\frac{d}{dt}(\eta_t^{-1}) = -T\eta_t^{-1} \circ \frac{d\eta}{dt} \circ \eta_t^{-1} .$$

The last formula follows from the chain rule applied to $\eta_t \circ \eta_t^{-1} = \text{id}$. □

So, putting this together, we get:

$$\frac{dv_t}{dt} = S(X_t) \circ \eta_t^{-1} - T(X_t \circ \eta_t^{-1}) \circ X_t \circ \eta_t^{-1}.$$

Now using the previous formula for $S(X)$, this becomes $= -(P_e \nabla_v v)_v^\ell$. Note especially the cancellation of the $Tv \circ v$ terms which has occurred. But as we recall $P_e(\nabla_{v_t} v_t) = \nabla_{v_t} v_t - \nabla p_t$ where p is a smooth function. We can identify $P_e(\nabla_{v_t} v_t)_v^\ell$ with $P_e(\nabla_{v_t} v_t)$ (since dv/dt really stands for its vertical lift) and hence get the Euler equations

$$\frac{dv_t}{dt} = -\nabla_{v_t} v_t + \nabla p_t$$

or

$$\frac{dv_t}{dt} + \nabla_{v_t} v_t = \nabla p_t.$$

(The minus sign on the pressure can be recovered by using $-p_t$.) Thus we have proved:

<u>Theorem.</u> <u>If X_t is an integral curve of the spray on $\mathcal{D}_\mu^s$, its pullback $v_t = X_t \circ \eta_t^{-1}$ does satisfy the Euler equations. In other words, η_t is a geodesic on $\mathcal{D}_\mu^s$ iff its velocity field satisfies the Euler equations.</u>

By inspecting the above calculation it becomes clear where the derivative loss occurs. If X is an H^s vector field on M, we know $S(X)$ is an H^s vector field on TM. However it is the sum of two H^{s-1} vector fields on TM. The top derivatives cancel, but when this is pulled back to Eulerian coordinates one of these terms disappears, namely $TX_t \circ X$ and so what we are left with is one of the H^{s-1} summands.

All of the above goes through for manifolds with boundary since the Hodge theorem projects vector fields at the boundary onto those which are tangent to the boundary as mentioned before.

As a consequence of these calculations we have this theorem, mentioned earlier.

<u>Theorem.</u> <u>Given</u> $v_0 \in T_e \mathcal{D}_\mu^s$ <u>there is an</u> $\epsilon > 0$ <u>and a unique vector field</u> $v(t) \in T_e \mathcal{D}_\mu^s$ <u>for</u> $-\epsilon < t < \epsilon$ <u>which satisfies the Euler equation. Moreover, these solutions</u> v_t <u>depend continuously on the initial data</u> v_0.

<u>Proof.</u> For the existence part of the theorem it is sufficient to find short-time solutions to the geodesic spray on $\mathcal{D}_\mu^s(M)$. But since $\mathcal{D}_\mu^s(M)$ is a Hilbert manifold and the spray is smooth the existence follows immediately from the existence theorem for C^r vector fields on Banach manifolds (see lecture 1).

The continuous dependence on initial conditions follows from the fact that the pullback $v_t = X_t \circ \eta_t^{-1}$ involves left composition so it is continuous (but not smooth). The initial condition for the spray on $\mathcal{D}_\mu^s(M)$ is an element of $T_e \mathcal{D}_\mu^s(M)$ since we are interested in flows in $\mathcal{D}_\mu^s(M)$ starting at the identity. □

This existence theorem has been proved in weaker forms by

Lichtenstein [1] and Guynter [1]. The general case of manifolds with boundary is due to Ebin-Marsden [1].

The flow in Lagrangian coordinates is C^∞. In Euler coordinates, let $E_t(v_0) = v_t$ be the solution flow. Then for <u>fixed</u> t, E_t is a continuous map, but is probably not differentiable.*
Various smoothness properties of the Euler and Navier-Stokes equations are important in developments discussed in the next lecture (see Marsden [7]).

The proof of the above Theorem is based on the existence of integral curves for the spray S. This in turn follows from the fundamental existence theorem for ordinary differential equations. Recall that this theorem is proven by showing an iteration (called Picard iteration) always yields solutions. So, by inspecting the above proof it should be possible to find an approximation procedure which converges to solutions.

This in fact, points out an essential difference between working with the whole spray and working with its pullback $P_e(\nabla_v v)$. The Picard method will not in general converge for the Euler equation as it stands. Indeed in practical numerical computations, one often uses Lagrangian coordinates. (See also the next lecture.)

---

* Indeed, Kato [5] has shown that the evolution operator $U_t : H^s \to H^s$ for $\frac{\partial u}{\partial t} + u\frac{\partial u}{\partial x} = 0$ on R is continuous, but is not Holder continuous for any exponent α, $0 < \alpha \leq 1$.

5. Turbulence and Chorin's Formula.

This lecture is concerned with some aspects of the Navier Stokes equations which are connected with turbulence. We shall be beginning with a representation theorem for the solution of the Navier-Stokes equations which was discovered by A. Chorin in an attempt to find a good numerical scheme to calculate solutions. This scheme is important in that it allows good calculations at interesting Reynolds numbers. One writes the Navier-Stokes equations as

$$\begin{cases} \dfrac{\partial v}{\partial t} - \dfrac{1}{R} \Delta v + (v \cdot \nabla) v = -\text{grad } p \\ \text{div } v = 0 \\ v = 0 \quad \text{on} \quad \partial M \end{cases}$$

and calls $R = 1/(\text{viscosity})$ the Reynolds number; if one rescales v to Vv, distances by a factor d and time by d/V we get a new solution with $R = Vd/\nu$. Most numerical schemes break down with R a few hundred, but Chorin's scheme is valid far beyond that, possibly up to $R = 50,000$. Our goal is to present the formula and to discuss where it comes from and its plausibility. The second part of the lecture will discuss some aspects of turbulence theory. This subject is basically concerned with qualitative features of the solutions as $R \to \infty$. The approach here follows that of Ruelle-Takens [1].

Statement of Chorin's Formula.

Let us write the Navier-Stokes equations as follows:

$$\begin{cases} \dfrac{\partial v}{\partial t} = \widetilde{\Delta} v + Z(v) \\ v = 0 \quad \text{on} \quad \partial M \end{cases}$$

where $\tilde{\Delta} = \frac{1}{R} P \cdot \Delta$ and $Z(v) = -P((v \cdot \vec{\nabla})v)$. Here P is the projection onto the divergence free part discussed in the last lecture (Δv is divergence free, but need not be parallel to ∂M, so one still requires a P in front of Δv).

Let H_t denote the evolution operator or semi-group defined by $\tilde{\Delta}$. It exists because it is an elementary exercise to show that $\tilde{\Delta}$ is self adjoint and ≤ 0 on the Hilbert space $L_2(M)$ with domain $H_0^2(M)$. (See the parabolic form of the Hille-Yosida theorem discussed in lecture 1). Thus H_t is defined for $t \geq 0$, and solves $\partial v/\partial t = \tilde{\Delta} v$. (This is called the "Stokes" equation.)

Let E_t denote the evolution operator for the Euler equations which was obtained in the last lecture.

Let F_t denote the full solution to the Navier-Stokes equations.

Let $\varphi(v)$ be a potential for v; e.g.: $\varphi(v) = d\Delta^{-1}(v)$, so $v = \delta(\varphi(v))$. Here δ is the divergence operator discussed in lecture 3. (More concretely in three dimensions, $v = \vec{\nabla} \times \varphi(v)$.) Let $d(\ell)$ be a function of $\ell \in R$, $\ell \geq 0$ with $d(\ell) = \sqrt{\ell \nu}$ where $\nu = 1/R$ is the viscosity of the fluid. It will turn out that $d(\ell)$ will be a measure of the thickness of the boundary layer.

Let g_ℓ be a C^∞ function equal to one a distance $\geq d(\ell)$ from ∂M and $g_\ell = 0$ on a neighborhood of ∂M.

Define the operator

$$\Phi_\ell(v) = \delta(g_\ell \cdot \varphi(v)),$$

we call Φ_ℓ the <u>vorticity creation operator</u>. The reason for this is that $\Phi_\ell(v)$ equals v away from ∂M, but if v is only $\|\partial M$, $\Phi_\ell(v)$ will be zero on ∂M so has the effect of "chopping off" v within the boundary layer (we do not use $g_\ell \cdot v$ since that is not divergence free). Such a chopping off effectively creates vorticity. (See the figure following.)

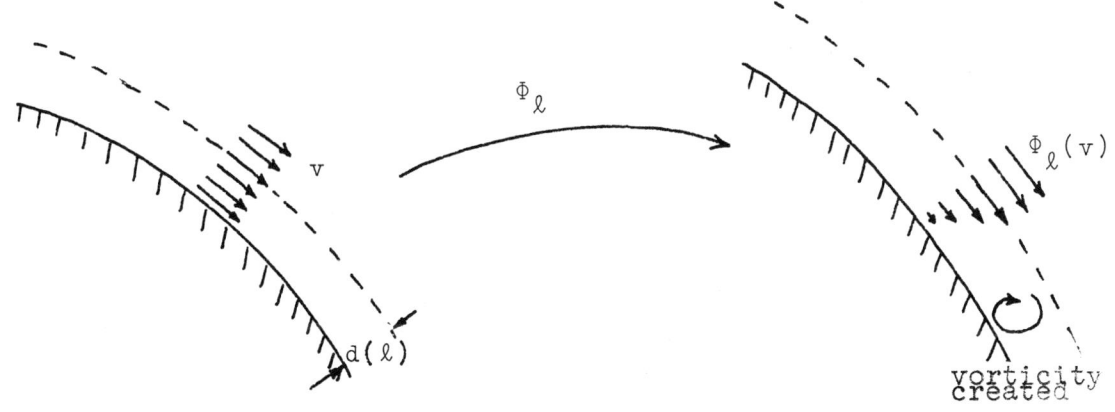

The formula now reads as follows:

$$F_t(v) = \text{solution of Navier-Stokes equation}$$
$$= \lim_{n \to \infty} (H_{t/n} \circ \Phi_{t/n} \circ E_{t/n})^n v.$$

In this formula the power means iteration. For example:

$$(H_{t/3} \circ \Phi_{t/3} \circ E_{t/3})^3 v = H_{t/3} \circ \Phi_{t/3} \circ E_{t/3} \circ H_{t/3} \circ \Phi_{t/3} \circ E_{t/3} \circ H_{t/3} \circ \Phi_{t/3} \circ E_{t/3} \cdot v.$$

Thus one divides the time scale into n parts and then iterates the

the procedure: solve Euler's equations then create vorticity, then solve the Stokes equation then the Euler equation, etc.

This is the basic method underlying Chorin's technique. However part of the beauty of the method is the way in which he solves numerically for E_t and H_t. He uses vorticity methods for E_t and probabilistic methods for H_t. See Chorin [2] for details.

In the following figure* we reproduce one of Chorin's outputs. The 0's mark negative vorticity and *'s mark positive vorticity. This representation is for flow past a cylinder with R and t as marked and initial v corresponding to parallel flow. It is a remarkable achievement to obtain on the computer something resembling the famous "Karmen vortex street". (For a spectacular photograph, see Scientific American, January 1970, p. 40; this is reproduced on the cover of "Basic Complex Analysis", W. H. Freeman Co. (1973).) Below we shall discuss further the qualitative features of why and how such periodic phenomena can get generated.

As is well known (Nelson [3]) product formulas are closely related to Wiener integrals. Chorin has recently used this idea to improve the scheme still further, as far as computer efficiency goes, so the method is valid into the fully turbulent region.

---

* The computer has distorted the cylinder somewhat into an ellipse.

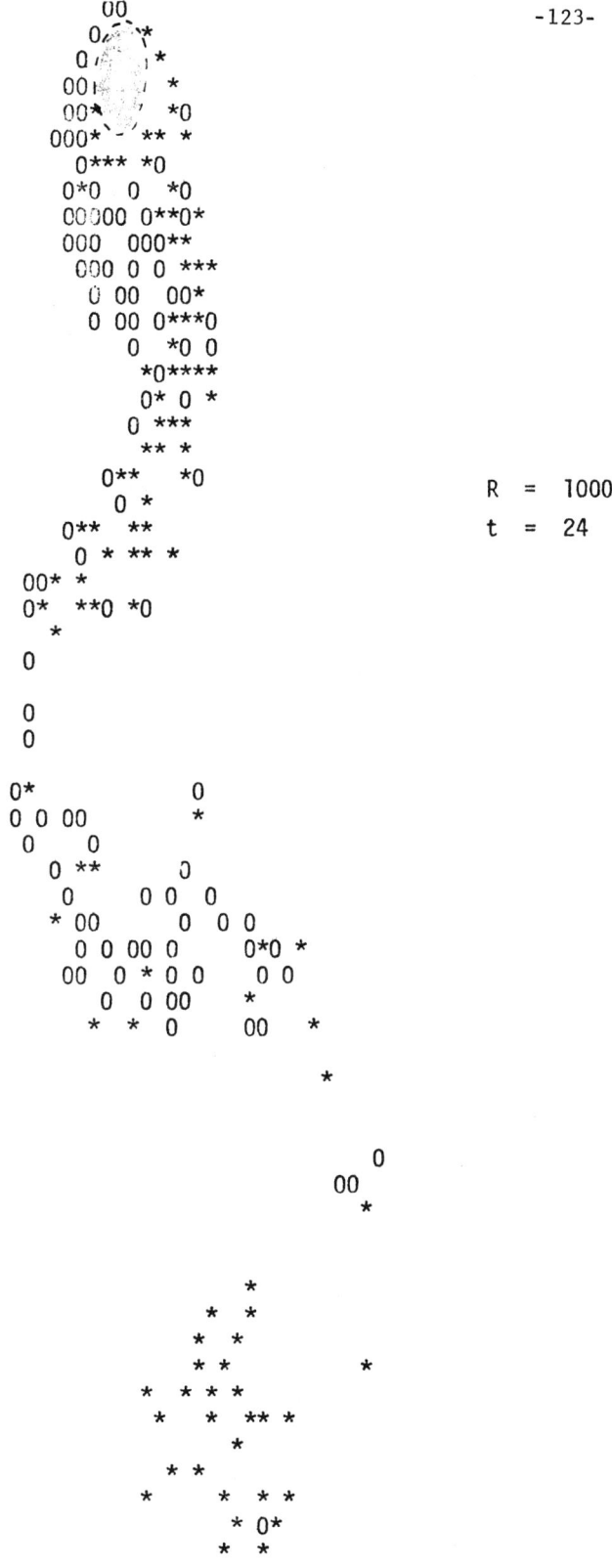

R = 1000
t = 24

The interesting feature of the above formula is that the error for large n is $O(1/n)$ independent of R. Furthermore using the formula as an existence theorem we find that smooth solutions to the Navier-Stokes equations exist for a time interval $T > 0$ independent of R as $R \to \infty$ and converge in L_p to solutions of the Euler equations.

This is an important result, for it guarantees as positive time of existence for given initial data, no matter how small the viscosity. This is strong evidence for the existence of smooth turbulent solutions. (See below.)

In case $\partial M = \emptyset$ (for example using periodic boundary conditions) the formula reads

$$F_t v = \lim_{n \to \infty} (H_{t/n} \circ E_{t/n})^n v .$$

This formula was proven in Ebin-Marsden [1] and Marsden [5]. It enabled us to show that as $\nu \to 0$ (or $R \to \infty$) the solutions converge in H^s to solutions of the Euler equations. (See also Swann [1], Kato [2].) Basically this means that turbulence cannot occur if no boundaries are present. Such convergence will not occur if $\partial M \neq \emptyset$ in topologies stronger than L_p because the boundary conditions and the vorticity carried into the mainstream flow will not allow it.

The complete proofs of these results are too technical for us to go into here. Rather we shall confine ourselves, in the next section, to an elementary exposition of where these formulas come from. We shall also include some additional intuition below.

The Lie-Trotter Formula.

Let X and Y be vector fields with flows H_t and E_t. Then the flow F_t of $X + Y$ is given by

$$F_t = \lim_{n \to \infty} (H_{t/n} \circ E_{t/n})^n.$$

Theorem. *This is valid if* X, Y *are* C^r *vector fields for those* t *for which* F_t *is defined.*

Let us give the idea (for details, see e.g., Nelson [1]). We first show F_t defined by the limit is a flow. One shows $F_{t+s} = F_t \circ F_s$ first if s, t are rationally related and takes limits. Consider, e.g.: $t = s$.

$$F_{2t} = \lim_{n \to \infty} (H_{2t/n} \circ E_{2t/n})^n$$

$$= \lim_{n \to \infty} (H_{2t/2n} \circ E_{2t/2n})^{2n}$$

$$= \lim_{n \to \infty} (H_{t/n} \circ E_{t/n})^{2n}$$

$$= \lim_{n \to \infty} (H_{t/n} \circ E_{t/n})^n \circ (H_{t/n} \circ E_{t/n})^n$$

$$= F_t \circ F_t.$$

Next one shows $\frac{d}{dt} F_t(x) \big|_{t=0} = X(x) + Y(x)$. Indeed, formally,

$$\frac{d}{dt} F_t(x) \big|_{t=0} = \lim_{n \to \infty} \frac{d}{dt} (H_{t/n} \circ E_{t/n})^n x \big|_{t=0}$$

$$= \lim_{n \to \infty} \frac{d}{dt}(H_{t/n} \circ E_{t/n} \circ H_{t/n} \circ \cdots \circ H_{t/n} \circ E_{t/n}) x \big|_{t=0}$$

$$= \lim_{n \to \infty} [\frac{1}{n}(X(x)+Y(x))+\ldots+ \frac{1}{n}(X(x)+Y(x))]$$

$$= X(x) + Y(x) .$$

It follows now that F_t is the flow of $X+Y$ since

$$\frac{d}{dt}F_t(x) = \frac{d}{ds}F_{s+t}(x)\big|_{s=0}$$

$$= \frac{d}{ds}F_s(F_t(x))\big|_{s=0}$$

$$= X(F_t(x)) + Y(F_t(x)) .$$

The above formula arose historically in Lie group theory. It tells us how to exponentiate the sum of two elements in the Lie algebra. In the case of matrix groups it is the classical formula:

$$e^{t(A+B)} = \lim_{n \to \infty} (e^{tA/n} e^{tB/n})^n .$$

Of course if $[X, Y] = 0$ the formula reads $F_t = H_t \circ E_t$, but it really is the case in which X, Y do not commute that is of interest.

The above formula has been generalized to linear evolution equations, as in the Hille-Yosida theorem by Trotter [1], and to certain non-linear semi-groups by Brezis-Pazy [1] and Marsden [5]. These results can be used to establish the claims made about the Navier Stokes equation if $\partial M = \emptyset$. Indeed one takes $X = \tilde{\Delta}$ and $Y = Z$.

For $\partial M \neq \emptyset$ the composition $H_t \circ E_t$ doesn't even make sense (except perhaps in $L_2(M)$, but that is not too useful) because $E_t(v)$, even if $v = 0$ on ∂M, will not be 0 on ∂M, but will only be parallel to ∂M. The purpose of the vorticity creation operator is to correct for this failure of the boundary conditions.

Some additional intuition on Chorin's Formula.

Consider again the formula

$$F_t(v) = \lim_{n \to \infty} (H_{t/n} \circ \Phi_{t/n} \circ E_{t/n})^n v .$$

The term $E_{t/n} v$ gives the main overall features of the flow past the boundary. Let us call it the <u>downstream drift</u>. Consider the effect: $E_{t/n}$ drifts us downstream, then $\Phi_{t/n}$ creates vorticity near ∂M, then $H_{t/n}$ has the effect of diffusing this vorticity away from ∂M then $E_{t/n}$ tends to sweep this vorticity downstream etc. The net effect is a lot of vorticity swept downstream. This is exactly what happens in examples such as the von Karmen vortex street.

The proof of Chorin's formula is based on a generalization of the Lie Trotter product formula due to Chernoff [1] in the linear case and Brezis-Pazy [1] and Marsden [5] in the non-linear case. We discuss this formula next.

Chernoff's Formula.

Suppose $K(t)$ is a family of operators, $t \geq 0$ (satisfying suitable hypotheses). Let $X = K'(0)$. Then the flow of X is

$$F_t(x) = \lim_{n \to \infty} [K(t/n)]^n (x).$$

This is Chernoff's generalization of the Lie-Trotter formula. We obtain the previous formula for $X+Y$ using $K(t) = H_t \circ E_t$.

For details on the hypotheses, see the aformentioned references and Chernoff-Marsden [1] and Nelson [1].

In applications to hydrodynamics it is important to use Lagrangian coordinates, for as we have stressed in the previous lecture, the Euler equations then become a C^∞ vector field. This is a great advantage in dealing with these product formulas (in the linear case it corresponds to adding a bounded operator to an unbounded one -- a relatively easy procedure).

For example one can give an almost trivial proof of the formula

$$E_t = \lim_{n \to \infty} (P\mathring{E}_{t/n})^n$$

where $\mathring{E}_t$ is the evolution operator for $\frac{\partial u}{\partial t} + (u \cdot \nabla)u = 0$ whose solution is known explicitly. A similar theorem proved using Euler coordinates and with more effort was done by Chorin [1].

To obtain Chorin's formula as previously described, one chooses $K(t) = H_t \circ \Phi_t \circ E_t$.

Calculation of the Generator.

Probably the most crucial thing in Chorin's formula is the formal reason why $K'(0) = \tilde{\Delta} + Z$. Indeed we claim that Φ_t contributes nothing to $K'(0)$. This is, of course, crucial if our resulting flow is to be associated with the Navier-Stokes equations. In the following we attempt to show why $K'(0) = \tilde{\Delta} + Z$ with $K(t)$ as above.

In order to see this, write

$$\frac{1}{t}\{H_t\Phi_t E_t v - v\} = \frac{1}{t}\{[H_t\Phi_t E_t v - H_t\Phi_t v] + [H_t\Phi_t v - H_t v] + [H_t v - v]\}.$$

The first and last terms converge, respectively to $Z(v)$ and $\tilde{\Delta}v$ (one needs to know $H_t\Phi_t$ is t-continuous for this). Thus the validity is assured by the following key lemma: if v is suitably smooth, $v = 0$ on ∂M, then in L_p,

$$\underset{t \to 0}{\text{limit}} \; \frac{1}{t}[H_t\Phi_t v - H_t v] = 0.$$

Indeed, if $K(t, x, y)$ is a Green's function for $\tilde{\Delta}$ on M then

$$\frac{1}{t}(H_t\Phi_t v - H_t v)(x) = \frac{1}{t}\int_M K(t, x, y)[(\Phi_t v)(y) - v(y)]dy$$

$$= \int_M \frac{1}{t} dK(t, x, y)[g_t\varphi(v)(y) - \varphi(v)(y)]dy$$

$$= \int_{B_t} \frac{1}{t} dK(t, x, y)[g_t\varphi(v)y - \varphi(v)(y)]dy$$

where $B_t = \{x \in M \mid d(x, \partial M) \leq d(t)\}$. Taking into account the nature of the singularity of K and the choice of $d(t)$ it is easy to see that in L_p norm, the above is majorized by

$$C_t \times \sup_M [g_t \varphi(v) - \varphi(v)]$$

where C_t, the L_p norm of $\frac{1}{t} \int_{B_t} dK(t, x, y) dy$ goes to zero as $t \to 0$ on account of the rapidity with which the volume of B_t goes to zero as $t \to 0$. This gives the formula.

The Hopf Bifurcation.

We now turn our attention to the qualitative nature of turbulence. Actually the literature is very confusing -- a few representative works are listed in the bibliography. However we wish to describe a theory due to Ruelle-Takens [1] which has several very attractive features.

Basically we want to study the Navier-Stokes equations and let $R \to \infty$. Thus we are interested in studying dynamical systems depending on a parameter. One of the most basic results in this regard is a theorem of Hopf from 1942 (Hopf [1]).

In order to understand Hopf's theorem, let us review some standard material in ordinary differential equations. For a complete discussion of this material, see Coddington-Levinson [1] and Abraham-Robbin [1]. Let $X : R^n \to R^n$ be a linear map. Then regarding X as a vector field on R^n, its flow is given by $F_t(a) = e^{tX}(a)$, where

$a \in \mathbb{R}^n$ and $e^{tX} = \sum_{n=0}^{\infty} (t^n X^n/n!)$; in this expression $X^0 = I$ and multiplication is as matrices. Let $\lambda_1, \ldots, \lambda_k$ be the (possibly complex) eigenvalues of X. Since X has only real entries when considered as a matrix, the λ_i appear in conjugate pairs. Clearly $e^{t\lambda_1}, \ldots, e^{t\lambda_k}$ are the eigenvalues of F_t.

Now suppose that for all i, we have $\text{Re}(\lambda_i) < 0$. Then as t increases $|e^{\lambda_i t}|$ is decreasing and hence the orbit of a point $a \in \mathbb{R}^n$ i.e., the curve $t \mapsto F_t(a)$, is approaching zero. (This is clear if X is diagonalizable; for the general case one uses the Jordan canonical form.) Since F_t is linear, for each t we have $F_t(0) = 0$. In this situation, we say 0 is an <u>attracting</u> or <u>stable fixed point</u>.

Now if all $\text{Re}(\lambda_i) > 0$, it is clear that each $|e^{t\lambda_i}|$ is increasing with t, and so the orbit of a point under the flow is away from 0. Here, we say 0 is a <u>repelling</u> or <u>unstable fixed point</u>.

For the nonlinear case, we linearize and apply the above results as follows. Let X be a vector field on some manifold M. Suppose there is a point $m_0 \in M$ such that $X(m_0) = 0$. Then F_t, the flow of X leaves m_0 fixed; $F_t(m_0) = m_0$. It makes sense to consider $DX(m_0) : T_{m_0} M \to T_{m_0} M$. If $y_1, \ldots, y_n$ is a coordinate system for M at m_0, the coordinate matrix expression for $DX(m_0)$ is just $DX(m_0) = (\partial x^i / \partial y^j)(m_0)$. Now, $DX(m_0)$ can be treated as a linear map on $\mathbb{R}^n$ and the same analysis as above applies. Hence m_0 is an attracting or repelling fixed point (or neither) for the flow of X depending on the sign of the real part of the eigenvalues of

$(\partial x^i/\partial y^j)(m_0)$. However if m_0 is attracting (when the real parts of the eigenvalues are < 0), it is only <u>nearby</u> points which $\to m_0$ as $t \to \infty$.

To begin our study of the Hopf theorem, let us consider a physical example of the general phenomenon of <u>bifurcation</u>. The idea in each case is that the system depends on some real parameter, and the system undergoes a sudden qualitative change as the parameter crosses some critical point. (For research in a slightly different direction and for more examples, consult the papers in Antman-Keller [1] and Zarantonello [1].)

<u>Example</u> . (Couette Flow). Suppose we have a viscous fluid between two concentric cylinders (see the following figure). Suppose further we forcibly rotate the cylinders in opposite directions at some constant angular velocity ρ which is our parameter. For ρ near 0, we get a steady horizontal laminar flow in the fluid. However as ρ reaches some critical point, the fluid breaks up into what are called <u>Taylor</u> <u>cells</u> and the fluid moves radially in cells from the inner cylinder to the outer one and vice versa. Note, that the directions of flow are such that flow is continuous.

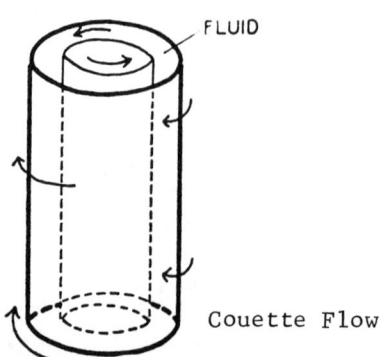
Couette Flow

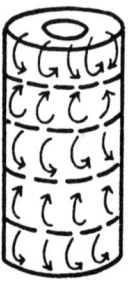
Taylor Cells

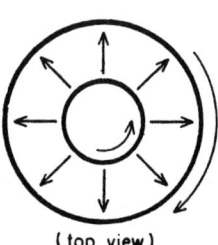
(top view)

In the above example, we have a situation described by differential equations and at some critical point of the parameter, the given solution becomes unstable and the system shifts to a "stable" solution. This sharp division of solutions is the sort of bifurcation we shall encounter in Hopf's theorem.

For simplicity, let us consider the case where the underlying space is simply R^2. Let X_μ be a vector field on R^2 depending smoothly on some real parameter μ. Actually it is convenient to put X_μ in R^3 by considering the map $\tilde{X} : (x, y, \mu) \mapsto (X_\mu(x, y), 0)$. This way we can graph the flow F_t^μ of X_μ and keep track of the parameter μ. The flow G_t of $\tilde{X}$ is $G_t(x, y, \mu) = (F_t^\mu(x, y), \mu)$. Similarly, we consider X_μ acting on the plane $\mu = $ const.

Now suppose $X_\mu(0, 0) = (0, 0)$ for each μ; more generally one could consider a curve (x_μ, y_μ) of critical points of X_μ. We can apply the analysis we developed for vector fields, i.e., for each μ, we look at the eigenvalues of $DX_\mu(0, 0)$ say $\lambda(\mu)$ and $\overline{\lambda(\mu)}$. (They are complex conjugate.) Note that the eigenvalues depend on μ and by our earlier analysis of flows, we know the qualitative behaviour of the flow depends on the sign of $\text{Re}(\lambda(\mu))$ and $\text{Re}(\overline{\lambda(\mu)})$ (which are equal in case $\lambda(\mu)$ itself is not real). So if we know how $\lambda(\mu)$ depends on μ then we can hope to extract some information about the flow near $(0, 0)$ as μ increases. We make these hypotheses:

Suppose $\text{Re}(\lambda(\mu)) < 0$ for $\mu < 0$ and $\text{Re}(\lambda(0)) = 0$ and $\text{Re}(\lambda(\mu))$ is increasing as μ increases across 0. Also assume that

$\lambda(\mu)$ <u>is not real and that for</u> $\mu = 0$, $(0, 0)$ <u>is an attracting</u> <u>fixed point for</u> X
(perhaps with a weaker or slower attraction than when $\text{Re}(\lambda(\mu)) < 0$).

Now for $\mu < 0$, we know from the above that the flow is "stable," i.e., points near $(0, 0)$ are carried towards $(0, 0)$ by the flow, as is the case for $\mu = 0$ (only slower) by assumption. The surprising case is the behavior for $\mu > 0$.

<u>Theorem.</u>* (E. Hopf). <u>In the situation described above, there is a</u> <u>stable periodic orbit for</u> X_μ <u>when</u> $0 < \mu < \varepsilon$ <u>for some</u> $\varepsilon > 0$. <u>(Stable here means points near the periodic orbit will remain near</u> <u>and eventually be carried closer to the orbit by the flow.)</u>

So as in the example we get a qualitative change in the stable solutions as μ crosses 0 , from an attracting fixed point at $(0, 0)$ to a periodic solution away from $(0, 0)$.

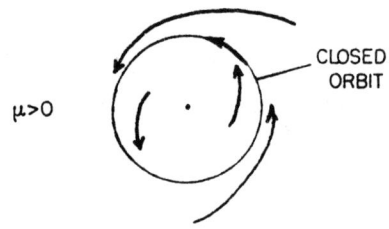

This theorem does generalize to R^n where we can get tori forming as the stable solutions (instead of closed orbits) as further bifurcations take place; see Ruelle-Takens [1] for details.

The proof of the theorem occurs in many places besides Hopf [1]. See, for instance Andronov and Chaikin [1], or Bruslinskaya

---

* See Ruelle [4] for a version suitable for systems with symmetry, such as Couette flow.

[3], or Ruelle-Takens [1].

Hopf's theorem is closely related to a linear model used in physics known as the "Turing model." As D. Ruelle, S. Smale, N. Kopell and H. Hartman have remarked, these sort of phenomena may be basic for understanding a large variety of qualitative changes which occur in nature, including biological and chemical systems. See for instance Turing [1], Selkov [1]. We have examined here only one of many types of possible bifurcations. There are many others which occur in Thom's theory of morphogenesis (see articles in Chillingworth [1] and Abraham [4] for more details and bibliography). Meyer [1] and Abraham [5] are representative of the Hamiltonian case.

For applications to fluid mechanics one wishes the vector field X_μ to be the Navier-Stokes equations and μ to be the Reynolds number. One is hampered by the fact that X_μ in this case is not a C^r vector field (even in Lagrangian coordinates). However this difficulty can be overcome and indeed the Hopf theorem is valid. For details see Marsden [3], Joseph-Sattinger [1], Iooss [1, 2], Judovich [3, 4], Bruslinskaya [1] etc.

Moreover, an important feature is that one can show that when a bifurcation does occur <u>one retains global existence of smooth solutions near the closed orbit</u>. This is in fact good evidence in the direction of verifying that the Navier-Stokes equations do <u>not</u> break down when turbulence develops.

Stability and Turbulence.

Shortly we shall explain more fully the Ruelle-Takens theory of turbulence. For now we just wish to stress the point that turbulence appears to be some complicated flow which sets in after successive bifurcations have occurred. In this process, stable solutions become unstable, as the Reynolds number is increased. Hence turbulence is supposed to be a necessary consequence of the equations and in fact of the "generic case" and just represents a complicated solution. For example in Couette flow as one increases the angular velocity Ω_1 of the inner cylinder one finds a shift from laminar flow to Taylor cells or related patterns at some bifurcation value of Ω_1. Eventually turbulence sets in. In this scheme, as has been realized for a long time, one first looks for a stability theorem and for when stability fails (Hopf [4], Chandresekar [1], Lin [1] etc.). For example, if one stayed closed enough to laminar flow, one would expect the flow to remain approximately laminar. Serrin [2] has a theorem of this sort which we present as an illustration:

<u>Stability Theorem.</u> <u>Let</u> $D \subset R^3$ <u>be a bounded domain and suppose the flow</u> v_t^ν <u>is prescribed on</u> ∂D <u>(this corresponds to having a moving boundary, as in Couette flow). Let</u> $V = \max_{\substack{x \in D \\ t \geq 0}} \|v_t^\nu(x)\|$, $d =$ <u>diameter of</u> <u>D and</u> ν <u>equal the viscosity. Then if the Reynolds number</u> $R = (Vd/\nu) \leq 5.71$, v_t^ν <u>is universally</u> L^2 <u>stable.</u>

Universally L^2 stable means that if $\bar{v}_t^\nu$ is <u>any</u> other solution to the equations and with the same boundary conditions, then the L^2 norm (or energy) of $\bar{v}_t^\nu - v_t^\nu$ goes to zero as $t \to 0$.

The proof is really very simple and we recommend reading Serrin [2] for the argument.

Chandresekar [1], Serrin [2], and Velte [1] have analyzed criteria of this sort in some detail for Couette flow.

As a special case, we recover something that we expect. Namely if $v_t^\nu = 0$ on ∂M is any solution for $\nu > 0$ then $v_t^\nu \to 0$ as $t \to \infty$ in L^2 norm, since the zero solution is universally stable.

Couette flow is not the only situation where this Taylor cell type of phenomenon occurs and where the above analysis is possible. For example, in the Bénard Problem one has a vessel of water heated from below. At a critical value of the temperature gradient, one observes convection currents, which behave like Taylor cells; cf. Rabinowitz [1].

This transition from laminar to periodic motion (the Hopf bifurcation) occurs in many other physical situations such as flow behind an obstacle.

A Definition of Turbulence.

A traditional definition (as in Hopf [2], Landau-Lifschitz [1]) says that turbulence develops when the vector field v_t can be described as $v_t(w_1, \ldots, w_n) = f(tw_1, \ldots, tw_n)$ where f is a quasi-periodic function, i.e., f is periodic in each coordinate, but the periods are not rationally related. For example, if the orbits of the v_t on the tori given by the Hopf theorem can be described by spirals with irrationally related angles, then v_t would such a flow.

Considering the above example a bit further, it should be clear there are many orbits that the v_t could follow which are qualitatively like the quasi-periodic ones but which fail themselves to be quasi-periodic. In fact a small neighborhood of a quasi-periodic function may fail to contain many other such functions. One might desire the functions describing turbulence to contain most functions and not only a sparse subset. More precisely, say a subset U of a topological space S is <u>generic</u> if it is a Baire set (i.e., the countable intersection of open dense subsets). It seems reasonable to expect that the functions describing turbulence should be generic, since turbulence is a common phenomena and the equations of flow are never exact. Thus we would want a theory of turbulence that would not be destroyed by adding on small perturbations to the equations of motion.

The above sort of reasoning lead Ruelle-Takens [1] to point out that since quasi-periodic functions are not generic, it is unlikely they "really" describe turbulence.[*] In its place, they propose the use of "strange attractors." (See Smale [2] and Williams [1].) These exhibit much of the qualitative behavior one would expect from "turbulent" solutions to the Navier-Stokes equations and they are stable under perturbations.

---

[*] See also Joseph-Sattinger [1].

Here is an example of a strange attractor. Let $U \subset R^n$ be open and $\sigma_t : U \to U$ some flow; suppose further for $x \in U$, there is an $s \in R$ such that $\sigma_{s+t}(x) = \sigma_t(x)$, i.e., x belongs to a periodic orbit of the flow. Let $(d/dt)\sigma_t(x)|_{t=0} = Y_x$ and let V be the affine hypersurface in U orthogonal to Y_x. For a small neighborhood S of x in V, there is a map $P : S \to V$ called the Poincaré map, defined as follows: For $w \in S$, it is easy to show there is a smallest $P_w \in R$ such that $\sigma_{P_w}(w) \in V$. Call $P(w) = \sigma_{P_w}(w)$. Now of course one can do this for each point of the periodic orbit. By doing this one gets a map on a small "tubular" neighborhood of the periodic orbit in U. (Here one must check that there is a neighborhood N os the orbit such that if $x \in N$ then x belongs to a unique hypersurface orthogonal to the orbit.) Also one can drop the condition that P be defined about a closed orbit by requiring that the vector field be almost parallel and everywhere transversal to a hypersurface V. In this case one can define a Poincaré map P over the entire space U by letting $P(x)$ be the first intersection of the integral curve through x with V.

In particular consider V to be a solid torus in three space and suppose we have a flow σ_t on U such that its Poincaré map wraps the torus around twice. Then the attracting set of the flow (i.e., $\{x \in U | x = \lim_{t \to \infty} \sigma_t(y)$ for some $y \in U\}$ is locally a Cantor set cross a 2-manifold (see Smale [2]). This is certainly a strange attractor! Ruelle-Takens [1] have shown if we define a strange attractor to be one which is neither a closed orbit or a point, then there are stable strange attractors on T^4 in the sense that a whole neighborhood of vector fields has a strange attractor as well.

If the attracting set of the flow, in the space of vector fields, which is generated by Navier-Stokes equations is strange, then a solution attracted to this set will clearly behave in a complicated, turbulent manner and since strange attractors are "generic", this sort of behavior should not be uncommon. Thus we have the following reasonable definition of turbulence as proposed by Ruelle-Takens:

"... the motion of a fluid system is turbulent when this motion is described by an integral curve of a vector field X_μ which tends to a set A , and A is neither empty nor a fixed point nor a closed orbit."

This turbulent motion is supposed to occur on one of the tori T^k that occurs in the Hopf bifurcation. This takes place after a <u>finite</u> number of successive bifurcations have occurred. However as S. Smale and C. Simon pointed out to us, there may be an infinite number of other qualitative changes which occur during this <u>onset of turbulence</u> (such as stable and unstable manifolds intersecting in various ways etc).

Since this sort of phenomena is supposed to be "generic," one would expect it to occur in other similar phenomena such as the Benard problem. (As the temperature gradient becomes very large, the flow becomes "turbulent.")

Recently Ruelle [1] (and unpublished work) has shown how the usual statistical mechanics of ergodic systems can be used to study the case of strange attractors, following work of Bowen [1] and Sinai [1]. It remains to connect this up with observed statistical properties of fluids, like the time average of the pressure in turbulent flow.

For the analytical nature of turbulent solutions, the work of Bass [1, 2] seems to be important.

In summary then, this view of turbulence may be phrased as follows. Our solutions for small μ (= Reynolds number in many fluid problems) are stable and as μ increases, these solutions become unstable at certain critical values of μ and the solution falls to a more complicated stable solution; eventually, after a certain finite number of such bifurcations, the solution falls to a strange attractor (in the space of all time dependent solutions to the problem). Such a solution, which is wandering close to a strange attractor, is called turbulent.

6. Symmetry Groups in Mechanics.

In this lecture we shall discuss the conservation laws resulting when one has a Hamiltonian system with symmetry. Intuitively one should think of linear and angular momentum which arise from translational and rotational invariance respectively. However one can have more sophisticated conservation laws too such as those dealing with spin, with the rigid body etc. Following these topics, we shall explain how one can shrink down the phase space in order to eliminate the variables which were obtained from the conservation laws. Some of these are subtle yet very fundamental, viz Jacobi's "elimination of the node" in celestial mechanics. Finally we shall discuss a completeness theorem in geometry and how various conservation laws can be used to prove it. Other completeness theorems are proved in lecture 8. This lecture is based on Souriau [1] and Marsden-Weinstein [1].

Before beginning the actual mechanics, we shall need a little notation and a few facts concerning Lie groups.

Preliminaries on Lie Groups and Group Actions.

Let G be a Lie group; i.e. a C^∞ manifold which is also a group and the group operations are C^∞. Let $\mathcal{G}$ denote the Lie algebra of G; we can think of $\mathcal{G}$ either as the vector space $T_e G$ or as the space of all left invariant vector fields on G. The latter gives us a bracket $[\xi, \eta]$ on $\mathcal{G}$ making it into a Lie algebra; i.e. $[[\xi,\eta],\zeta] [\zeta, \xi], \eta] + [\eta, \zeta], \xi] = 0$ holds.

Example 1. $SO(3)$ the group of all 3×3 orthogonal matrices of determinant +1 is a 3 dimensional Lie group. $\mathcal{G} = T_e SO(3)$ consists of the 3×3 skew adjoint matrices with bracket equalling the commutator. This space of matrices is, in turn identifiable with R^3. Making the identification, the bracket is just the cross product.

2. See the example $\mathcal{D}$, the diffeomorphism group, discussed in lecture 4.

By an <u>action</u> (or "non-linear representation") of G on a manifold M, we mean a collection of mappings $\Phi_g : M \to M$ such that

(i) $\Phi_{gh} = \Phi_g \circ \Phi_h$

and (ii) Φ_e = identity   e = identity in G.

We also require $(g, x) \mapsto \Phi_g(x)$ to be C^∞.

Notice that if $G = R$, an action is nothing more than a flow. As every flow determines a generating vector field, we are led to define the infinitesimal generators of an action. We do this in the following discussion.

Let $\xi \in \mathcal{G}$. Let $\exp \xi$ denote the exponential of ξ. (This is defined as follows; let $\tilde{\xi}$ denote the left invariant vector field which equals ξ at e; then $\exp t\xi$ is the integral curve of $\tilde{\xi}$ starting at e. For matrix groups $\exp \xi = e^A$ defined, e.g. as a power series.) Now one verifies $\exp t\xi$ is a one parameter subgroup of G; i.e., $\exp(t+s)\xi = \exp t\xi \cdot \exp s\xi$. Thus $\Phi_{\exp t\xi}$ is a flow on M. Let ξ_M denote its generator. We call the map $\xi \mapsto \xi_M$ of $\mathcal{G}$ to vector fields on M the <u>infinitesimal generator</u> of the action. One has $[\xi_M, \eta_M] = -[\xi, \eta]_M$.

Let G **act** on M and let $x \in M$. The <u>isotropy group</u> of x is, by definition:

$$G_x = \{g \in G | \Phi_g(x) = x\}.$$

It is a subgroup of G.

G is said to act <u>freely</u> on M if each $x \in M$ has $G_x = \{e\}$.

G is said to act <u>properly</u> on M if the map $(g, x) \mapsto (x, \Phi_g(x))$ of $G \times M \to M \times M$ is proper; i.e. inverse images of compact sets are compact.

If G acts on M and $x \in M$, $\{\Phi_g(x) | g \in G\} = G \cdot x$ is the <u>orbit</u> of x. These are always immersed submanifolds of M. (One maps $G/G_x \to G \cdot x$ to obtain the required immersion.) Moreover, M is the disjoint union of the orbits. Thus one can consider M/G the space of all orbits.

If G acts freely and properly on M then M/G is a C^∞ manifold and $\pi : M \to M/G$ is a submersion. Let $\pi(x) = [x]$. Now

$$T_x(G \cdot x) = \{\xi_M(x) | \xi \in \mathcal{G}\}$$

and

$$T_{[x]}(M/G) \cong T_xM/T_xG \cdot x.$$

(These facts are proven, for example in Bourbaki [1].) When we form quotient manifolds in the sequel we implicitly assume these hypotheses. If M consists only of one orbit, we say that we have a homogeneous space. Thus $M \cong G/G_x$. (In general M/G is not a manifold; consider S^1 acting on the plane; M/G is then a half ray.)

Let G act on M and on N by actions Φ_g and Ψ_g respectively. A map $\psi : M \to N$ is called equivariant (or an intertwining map) if

$$\psi \circ \Phi_g = \Psi_g \circ \psi \quad \text{for all } g \in G.$$

Consider now just a given Lie group G. Then there is an action of G on $\mathcal{G}$ by linear transformations called the adjoint action:

$$Ad_g \cdot \xi = TR_{g^{-1}} TL_g \cdot \xi.$$

Here R_g and L_g are the right and left translation maps. The infinitesimal generator of this action is $\xi \mapsto \xi_\mathcal{G}$, $\xi_\mathcal{G}(\zeta) = [\xi, \zeta] \equiv ad_\xi(\zeta)$.

We also get an action on the dual space $\mathcal{G}^*$ called the coadjoint action by using $(Ad_{g^{-1}})^*$.

The Moment Function of Souriau and Noethers Theorem.

We now consider a general setting for finding conservation laws. The basic results are due to Souriau [1] but were found also in Marsden [1] and Smale [4].

Definition. Let G be a Lie group and P a (weak) symplectic manifold. Let G act on P by symplectic diffeomorphisms. (It follows that each infinitesimal generator ξ_p satisfies $di_\xi \omega = 0$.)

By a <u>moment</u> for the action we mean a C^∞ map $\psi : P \to \mathcal{G}*$ such that if $\hat{\psi}$ denotes the dual map from $\mathcal{G}$ to the space of smooth function on P, i.e. $\hat{\psi}(\xi)(p) = \psi(p) \cdot \xi$, we have

$$d(\hat{\psi}(\xi)) = i_{\xi_p} \omega$$

i.e., $\langle T_p \psi \cdot v, \xi \rangle = \omega_p(\xi_p(p), v)$ for $\xi \in \mathcal{G}$, $v \in T_p P$. In other words, each infinitesimal generator ξ_p has $\hat{\psi}(\xi)$ as a Hamiltonian function. A moment, if it exists, is defined up to an arbitrary additive constant in $\mathcal{G}*$.

<u>Theorem</u>. <u>Let $H : P \to \mathbb{R}$ be invariant under Φ, i.e., $H \circ \Phi_g = H$. Then ψ is a constant of the motion for X_H; i.e. if F_t is the flow of X_H, $\psi \circ F_t = \psi$.</u>

<u>Proof</u>. From $H \circ \Phi_g = H$ it follows that $H \circ \Phi_{\exp t\xi} = H$ and hence $L_{\xi_p} H = 0$. But this means $\{\hat{\psi}(\xi), H\} = 0$ so $\hat{\psi}(\xi)$ is a constant of the motion. □

In order to actually compute ψ we use:

<u>Theorem</u>. <u>Let G act symplectically on P. Assume $\omega = -d\theta$ and the action leaves θ invariant. Then $\psi(p) \cdot \xi = (i_{\xi_p} \theta)(p)$ and ψ is equivariant; i.e. $\psi \circ \Phi_g = (\text{Ad}_{g^{-1}})^* \circ \psi$.</u>

Proof. Since Φ_g leaves θ invariant, we have $L_{\xi_p} \theta = 0$. Hence

$$di_{\xi_p} \theta + i_{\xi_p} d\theta = 0$$

i.e., $\quad i_{\xi_p} \omega = di_{\xi_p} \theta$.

Hence we can choose $\hat{\psi}(\xi) = i_{\xi_p} \theta$ as required. We leave equivariance of this formula as an exercise. □

Let us specialize further to give an even more useful formula:

Theorem. <u>Let G act on M. Then the action lifts to one on T^*M preserving the canonical one form (this was essentially proved in lecture 2).</u> We have

$$\psi(\alpha) \cdot \xi = \alpha \cdot \xi_M(x) \quad , \quad \alpha \in T^*_x M \ .$$

This follows in a straightforward way from the previous theorem. The quantities are sometimes written $P(X)(\alpha_x) = \alpha_x \cdot X(x)$, X a vector field on M and called the momentum functions. Equivariance can be phrased infinitesimally in terms of the <u>commutation relations</u>:

$$P([X, Y]) = -\{P(X), P(Y)\} \ .$$

In examples of linear or angular momentum the conserved quantity ψ reduces to the usual expressions.

We can also specialize to TM with a given metric rather than to T^*M with the canonical symplectic structure.

Theorem. Let G act on M by isometries, where M is a given Riemannian manifold. Let $V : M \to R$ be invariant and let $H(v) = \frac{1}{2}\langle v, v\rangle + V(x)$, $v \in T_x M$. Then if ξ_M is an infinitesimal generator of the action, the function

$$\hat{\psi}(\xi)(v_x) = \langle v_x, \xi_M(x)\rangle$$

is a constant of the motion for X_H.

It is also useful to present a version for general Lagrangian systems. The classical Noether theorem is a special case. Although this follows from the above, we give a separate proof here. (See lecture 2 for a general discussion of Lagrangian systems, and an explanation of the notation FL.)

In this result, observe that we do allow for the possibility that L might be degenerate. The only special assumption needed on Z is that it exist and be second order.

Proposition. Let Z be a Lagrangian vector field for $L : TM \to R$ and suppose Z is a second order equation.

Let Φ_t be a one parameter group of diffeomorphisms of M generated by the vector field $Y : M \to TM$. Suppose that for each real number t, $L \circ T\Phi_t = L$. Then the function $P(Y) : TM \to R$, $P(Y)(v) = FL(v) \cdot Y$ is constant along integral curves of Z.

Proof. Let $v(t)$ be an integral curve for Z. Then we shall show that $(d/dt)\{P(Y)(v(t))\} = 0$. Indeed, in a coordinate chart, if $(u(t), v(t))$ is the integral curve,

$$\frac{d}{dt}\{FL(v(t))\cdot Y\} = \frac{d}{dt}\{D_2L(u(t), v(t))\cdot Y(u(t))\}$$

$$= D_1D_2L(u(t), v(t))\cdot Y(u(t))\cdot \dot{u}(t) + D_2D_2L(u(t), v(t))$$

$$\cdot Y(u(t))\cdot \dot{v}(t) + D_2L(u(t), v(t))\cdot DY(u(t))\cdot \dot{u}(t).$$

Now the condition that Z be the Lagrangian vector field of L means exactly that the first two terms equal $D_1L(u(t), v(t))\cdot Y(u(t))$ (see the results given in lecture 2.). However if we differentiate $L\circ T\Phi_t$ with respect to t we obtain for any point (u, v),

$$0 = \frac{d}{dt}L(\Phi_t(u), D\Phi_t(u)\cdot v)\Big|_{t=0}$$

$$= D_1L(u, v)\cdot Y(u) + D_2L(u, v)\cdot DY(u)\cdot v.$$

Comparing this with the above gives $(d/dt)\{FL(v)\cdot Y\} = 0$ and proves the assertion. □

The Reduced Phase Space.

As mentioned in the introduction, when one has a group of symmetries, it is a classical procedure to eliminate a number of variables in order to get rid of the symmetries. We present now, following Marsden-Weinstein [1] a unified, as well as simplified, scheme for carrying out such a program.

For simplicity we shall always assume the moment ψ is

equivalent with respect to the $Ad_{g^{-1}}^*$ action. (This is not really necessary, for Souriau has shown that one can suitably modify the action.)

Let μ be a regular value of ψ; i.e., ψ is a submersion on $\psi^{-1}(\mu)$, so $\psi^{-1}(\mu)$ is a submanifold.

Let G_μ be the isotropy group of μ for the G action on $\mathcal{G}^*$. By equivarience, $\psi^{-1}(\mu)$ is invariant under G_μ so the orbit space $\psi^{-1}(\mu)/G_\mu$ is defined. Note also that by equivariance if $p \in \psi^{-1}(\mu)$ and $\Phi_g(p) \in \psi^{-1}(\mu)$ then $g \in G_\mu$. We let

$$P_\mu = \psi^{-1}(\mu)/G_\mu$$

and call P_μ the <u>reduced phase space</u>.

The main result is as follows.

<u>Theorem.</u> <u>Let</u> G <u>be a Lie group acting symplectically on the symplectic manifold</u> P, ω. <u>Let</u> ψ <u>be a moment for the action.</u> <u>Let</u> $\mu \in \mathcal{G}^*$ <u>be a regular value of</u> ψ. <u>Suppose</u> G_μ <u>acts freely and properly on the manifold</u> $\psi^{-1}(\mu)$. <u>Then if</u> $i_\mu : \psi^{-1}(\mu) \to P$ <u>is inclusion, there is a unique symplectic structure</u> ω_μ <u>on the reduced phase space</u> P_μ <u>such that</u> $\pi_\mu^* \omega_\mu = i_\mu^* \omega$, <u>where</u> π_μ <u>is the projection of</u> $\psi^{-1}(\mu)$ <u>onto</u> P_μ.

To prove this we shall make use of the following:

<u>Lemma.</u> <u>For</u> $p \in \psi^{-1}(\mu)$ <u>we have</u>

(i) $T_p(G_\mu \cdot p) = T_p(G \cdot p) \cap T_p(\psi^{-1}(\mu))$

and (ii) $T_p(\psi^{-1}(\mu))$ is the ω-orthogonal complement of $T_p(G \cdot p)$.

Proof. (i) Let $\xi \in \mathcal{G}$, so $\xi_p(p) \in T_p(G \cdot p)$. We must show $\xi_p(p) \in T_p(\psi^{-1}(\mu))$ iff $\xi \in \mathcal{G}_\mu$ the Lie algebra of G_μ. Equivariance gives $T_p \psi \cdot \xi_p(p) = \xi_{\mathcal{G}^*}(\mu)$, so $\xi \in \mathcal{G}_\mu$ iff $\xi_{\mathcal{G}^*}(\mu) = 0$ iff $\xi_p(p) \in \ker T_p \psi = T_p(\psi^{-1}(\mu))$.

(ii) For $\xi \in \mathcal{G}$, $v \in T_p P$ we have $\omega(\xi_p(p), v) = \langle T_p \psi \cdot v, \xi \rangle$ since ψ is a moment. Thus $v \in \ker T_p \psi$ iff $\omega(\xi_p(p), v) = 0$ for all $\xi \in \mathcal{G}$. $\square$

In the following proof we use the fact that if $F \subset E$ is a subspace of a symplectic space E, then $(F^\perp)^\perp = F$ where $\perp$ is the ω-orthogonal complement. In finite dimensions this follows by dimension counting. It is also true in infinite dimensions for weak symplectic forms if E is reflexive. We now prove our theorem.

Proof. For $v \in T_p(\psi^{-1}(\mu))$, let $[v] \in T_{\pi_\mu(p)} P_\mu$ denote the corresponding equivalence class in $T_p \psi^{-1}(\mu)/T_p(G_\mu \cdot p)$, so $[v] = T\pi_\mu \cdot v$. The assertion $\pi_\mu^* \omega_\mu = i_\mu^* \omega$ becomes

$$\omega_\mu([v], [w]) = \omega(v, w), \text{ for all } v, w \in T_p \psi^{-1}(\mu).$$

Thus ω_μ is unique. Moreover, ω_μ is well-defined because of the lemma. Also ω_μ is smooth because quantities on a quotient M/G are smooth when they have smooth pull-backs to M. Thus ω_μ is a well-defined smooth two-form on P_μ.

To show ω_μ is symplectic we first show ω_μ is non-degenerate; $\omega_\mu([v], [w]) = 0$ for all $w \in T_p \psi^{-1}(\mu)$ implies $v \in T_p(G_\mu \cdot p)$ by the lemma, or $[v] = 0$. It remains to show ω_μ is closed. But from $\pi_\mu^* \omega_\mu = i_\mu^* \omega$ and $d\omega = 0$, we conclude that $\pi_\mu^*(d\omega_\mu) = 0$, so $d\omega_\mu = 0$ since $T\pi_\mu$ is surjective. □

Remarks. Even if $\omega = -d\theta$ and the action leaves θ invariant, ω_μ need not be exact. For $\mu \neq 0$, θ does not project to a one-form on P_μ because $\theta(\xi_P)(p) = \psi(p)\xi \neq 0$.

As a consequence, observe that (in the finite dimensional case) P_μ is even-dimensional. If ψ is a submersion, then $\dim P_\mu = \dim P - \dim G - \dim G_\mu$.

If μ is a regular value of ψ, the action is always locally free near $\psi^{-1}(\mu)$.

Examples. 1. Let us begin by recalling the cotangent bundle case. Namely, if G acts on a manifold M, we obtain a symplectic action on T^*M which preserves the canonical one-form θ on T^*M. A moment for this action is given by $\psi : T^*M \to \mathcal{G}^*$:

$$\langle \psi(\alpha), \xi \rangle = \langle \alpha, \xi_M(m) \rangle, \quad \alpha \in T_m^*M.$$

By an earlier general theorem, this moment is Ad*-equivarient.

We conclude that if G_μ acts freely and properly on

$\psi^{-1}(\mu) = \{\alpha \in T^*M | <\alpha, \xi_M(m)> = <\mu, \xi> \text{ for all } \xi \in \mathcal{G}\}$, then $\psi^{-1}(\mu)/G_\mu$ is a symplectic manifold. If the $\xi_M(m)$ span a space of dimension = dim $\mathcal{G}$ at m, then it is easy to see that each point of $T^*_m M$ is regular.

2. If we specialize example 1, taking M = G with G acting on itself by left multiplication, then the moment $\psi : T^*G \to \mathcal{G}^*$ is given by

$$\psi(\alpha) = (TR_g)^* \cdot \alpha \in T^*_e G = \mathcal{G}^*, \quad \alpha \in T_g G$$

where R_g denotes right translation (cf. Arnold [1], Marsden-Abraham [1]). Thus each $\mu \in \mathcal{G}^*$ is regular and $\psi^{-1}(\mu)$ is the graph of the right invariant one-form ω_μ whose value at e is μ. Now $G_\mu = \{g \in G | L_g^* \omega_\mu = \omega_\mu\}$, so the action of G_μ on $\psi^{-1}(\mu)$ is left translation on the base point. Thus $\psi^{-1}(\mu)/G_\mu \approx G/G_\mu \approx G \cdot \mu \subset \mathcal{G}^*$. Thus the reduced phase space is just the orbit of μ in $\mathcal{G}^*$. That this is a symplectic manifold then follows from the above theorem. The rather special construction in this case is due to Kirillov-Kostant; see Kostant [1]. If one traces through the definitions one finds for $\beta \in G \cdot \mu$, $\gamma_1 = (\text{ad} u_1)^*\beta$ and $\gamma_2 = (\text{ad} u_2)^*\beta$, that

$$\omega_\mu(\gamma_1, \gamma_2) = \beta([u_2, u_1]).$$

When viewed directly, the symplectic structure on $G \cdot \mu \subset \mathcal{G}^*$ seems rather special. However, it becomes natural when viewed in the context of reduced phase spaces. Moreover, the proof becomes more transparent. This example is studied further below.

3. If G acts on M and leaves a given closed two-form F on M invariant, then we get a symplectic action on T*M with the symplectic form $\omega_F = \omega + \pi^* F$ where ω is the canonical form and $\pi : T^*M \to M$ the projection. Such a situation arises when one has a particle moving in the "electromagnetic field" F (see Souriau [1] and Sniatycki-Tulczyjew [2]). Now suppose F = dA is exact and A is invariant. Then the moment is given by

$$\langle \psi(\alpha), \xi \rangle = \langle \alpha - A, \xi_M(m) \rangle$$

(this corresponds to the classical prescription of replacing p by $p - \frac{e}{c} A$ in an electromagnetic potential A). The verification is the same as in example 1. Thus again, if μ is a regular value and G_μ acts freely and properly on $\psi^{-1}(\mu)$, we can form the reduced phase space P_μ.

4. Let G = SO(3) and P a symplectic manifold. Here $\mathcal{G} \approx R^3$ and the adjoint action is the usual one. For $\mu \in R^3$, $\mu \neq 0$, $G_\mu = S^1$ corresponding to rotations about the axis μ. (Since G is semi-simple, a symplectic action of G on P has an Ad*-equivariant moment ψ by Souriau [1]). One refers to ψ as "angular momentum" in this case. The reduction of P to $\psi^{-1}(\mu)/S^1$ is a generalization of the procedure called "elimination of the nodes" (cf. Smale [4] and Whittaker [1, p. 344]).

5. Suppose we have the situation of the above theorem, and in addition G is abelian. Ad*-equivariance means that the generating

functions $\hat{\psi}(\xi)$ are all in involution on P. Furthermore, $G_\mu = G$ for each $\mu \in \mathcal{G}^*$. If the action is free and μ is a regular value, we can form $P_\mu = \psi^{-1}(\mu)/G$. In this case $\dim P_\mu = \dim P - 2 \dim G$. The reduction to P_μ represents the classical reduction of a Hamiltonian system by integrals in involution.

As a special case, let X_H be a Hamiltonian vector field on P, so that the flow of X_H yields an action of R on P. The moment is just H itself so we get a symplectic structure on $H^{-1}(e)/R$ which is just the space of orbits on each energy surface (we assume e is a regular value of H).

6. Let $\mathcal{D}$ denote the group of C^∞-diffeomorphisms of a finite dimensional Riemannian manifold M. Suppose M is compact, or restrict to diffeomorphisms which are "asymptotic to the identity". Now as we saw in lecture 4, $T_e \mathcal{D} = \mathfrak{X}(M) =$ the vector fields on M and we put on $\mathcal{D}$ the L_2 metric which is obtained from $\mathfrak{X}(M)$ by right invariance. Thus $\mathcal{D}$ acting on $T\mathcal{D}$ on the right is a symplectic action. As in example 2, we conclude that for each $X \in \mathfrak{X}(M)$, the set $\{\eta * X | \eta \in \mathcal{D}\} \subset \mathfrak{X}(M)$ is a weak symplectic manifold. The symplectic structure is

$$\omega_X(\eta * L_{Y_1} X, \eta * L_{Y_2} X) = \int_M \langle X, [Y_2, Y_1] \rangle dx .$$

One may similarly restrict to volume preserving diffeomorphisms and divergence free vector fields. This symplectic manifold is left invariant by the Euler equations on $\mathfrak{X}(M)$ and they define a Hamiltonian system so restricted. (See the following theorem and corollary).

7. Let M and $\mathcal{D}$ be as in example 6. Let $\mathfrak{M}$ denote the space of all Riemannian metrics on M. Define the DeWitt metric on $\mathfrak{M}$ by

$$G_g(h, k) = \int_M [\langle h, k \rangle - (\mathrm{tr}\, h)(\mathrm{tr}\, k)] d\mu_g$$

Where $h, k \in T_g\mathfrak{M}$ = the symmetric 2-tensors on M, $\langle h, k \rangle$ is the inner product of h, k using the metric g, tr denotes the trace, and μ_g is the volume element associated with g. G_g is a weak metric and gives a (weak) symplectic structure on $T\mathfrak{M}$.

The space $T\mathfrak{M}$ is a basic (weak) symplectic manifold used in general relativity. We will now describe its reduced phase space in the presence of the symmetry group $\mathcal{D}$. (See lecture 9 for the connections of these ideas with general relativity.) $\mathcal{D}$ acts symplectically on $T\mathfrak{M}$ by pull-back. The moment for this action is not difficult to compute. It is:

$$\psi(g, k) \cdot X = 2 \int_M \langle X, \delta\pi \rangle d\mu_g$$

where $\pi = k - \frac{1}{2}(\mathrm{tr}\, k)g$ and δ is the divergence taken with respect to g. Of particular interest is the case $\psi^{-1}(0) = \{(g, k) \in T\mathfrak{M} \mid \delta\pi = 0\}$ (referred to as the divergence constraint in general relativity).

The isotropy group is all of $\mathcal{D}$, so the reduced phase space is $\psi^{-1}(0)/\mathcal{D}$. If we work near a metric with no isometries (asymptotically the identity if M is not compact), then $\psi^{-1}(0)/\mathcal{D}$ is a manifold by using methods explained in lecture 10. We conclude that $\psi^{-1}(0)/\mathcal{D}$ is a (weak) symplectic manifold.* This is the basic space one uses for a dynamical formulation of general relativity. It is related to "superspace" $\mathcal{M}/\mathcal{D}$ in that all "geometrically equivalent" objects have been identified. See Marsden-Fischer [1] for further results along these lines.

8. Let $\mathcal{H}$ be complex Hilbert space with $\omega = \text{Im}\langle,\rangle$ and $G = S^1$. Then G acts symplectically on $\mathcal{H}$ by $\Phi_z(\varphi) = z \cdot \varphi$, $|z| = 1$, $\varphi \in \mathcal{H}$. A moment is easily seen to be

$$\psi(\varphi) \cdot z = \tfrac{1}{2}\langle\varphi, \varphi\rangle \cdot z .$$

Thus $\psi^{-1}(1)$ is the unit sphere, so $\psi^{-1}(1)/G$ is projective Hilbert space. We recover the well-known fact that projective Hilbert space is a symplectic manifold (in fact it has a Kahler structure). This result will be useful for the next lecture.

Hamiltonian Systems on the Reduced Phase Space.

Theorem. *Let the conditions of the above theorem hold. Let K be another group acting symplectically on P with a moment φ. Let the actions of K and G commute and φ be invariant under G. Then*

(i) *K leaves ψ invariant*

(ii) *the induced action of K on P_μ is symplectic and has a moment which is naturally induced from the moment φ.*

---

* It is a conjecture of D. Ebin that this is true globally.

Proof. (i) This follows as in the proof that ψ is conserved by any G invariant Hamiltonian system on P (see above).

To prove (ii), let Ψ_k denote the action of $k \in K$ on P. By (i), $\psi^{-1}(\mu)$ is invariant under this action, and since the action commutes with that of G, we get a well-defined action on P_μ. Also, if $\tilde{\Psi}_k$ is the induced action on P_μ,

$$\pi_\mu^* \tilde{\Psi}_k^* \omega_\mu = \Psi_k^* \pi_\mu^* \omega_\mu = \Psi_k^* i^* \omega = i^* \Psi_k^* \omega = i^* \omega.$$

Hence $\tilde{\Psi}_k^* \omega_\mu = \omega_\mu$. Similarly, from the definition of moment we see that the induced moment is a moment for the induced action: namely, the induced moment $\tilde{\varphi}$ satisfies $\tilde{\varphi} \circ \pi_\mu = \varphi$, so for $[v] = T\pi_\mu \cdot v \in TP_\mu$, $\xi \in \mathcal{G}_K$, we have

$$\langle T\tilde{\varphi} \cdot [v], \xi \rangle = \langle T\varphi \cdot v, \xi \rangle = \omega(\xi_P, v) = \omega_\mu(\xi_{P_\mu}, [v])$$

since, as is easy to see, the generators ξ_P, ξ_{P_μ} on $\psi^{-1}(\mu)$ and P_μ are related by the projection π_μ. $\square$

For example, if we consider example 2 and let $G = K$ acting on T^*G by lifting the <u>right</u> action, we can conclude that the natural action of G on the orbit $G \cdot \mu \subset \mathcal{G}^*$ is a symplectic action. The induced moment is easily seen to be just the identity map: $\tilde{\varphi}(Ad^*_g \mu) = Ad^*_g \mu \in \mathcal{G}^*$.

The fact that G acts symplectically on the orbit $G \cdot \mu$, so that $G \cdot \mu$ is a "homogeneous Hamiltonian G-space", is a known and useful result. See Kostant [1] and Souriau [1, p. 116].

Taking $K = R$, we are led to:

Corollary. <u>Let the conditions of the theorem preceeding the above hold and let X_H be a Hamiltonian vector field on P with H invariant under the action of G. Then the flow of X_H induces a Hamiltonian flow on P_μ whose energy $\tilde{H}$ is that induced from H; i.e.,</u>
$\tilde{H} \circ \pi_\mu = H \circ i_\mu$.

For example if $\langle \, , \, \rangle$ is a left invariant metric on a group G, the Hamiltonian $H(v) = \frac{1}{2}\langle v, v \rangle$, which yields geodesics on G, induces a Hamiltonian system on the orbits in $\mathcal{G}^* \approx \mathcal{G}$. Note that the original Hamiltonian system on P is completely determined by the induced systems on the reduced spaces P_μ.

Similarly, in each of the other examples above, if we start with a given Hamiltonian system on P, invariant under G, then we can, with no essential loss of information, pass to the Hamiltonian system on the reduced phase space.

Relative Equilibria and Relative Periodic Points.

Definition. In the situation of the above corollary, a point $p \in P$ such that $\pi_\mu(p) \in P_\mu$ is a critical point [resp. periodic point] for the induced Hamiltonian system on P_μ is called a <u>relative equilibrium</u> [resp. <u>relative periodic point</u>] of the original system.

Poincaré [1] considered relative periodic points in the n-body problem on an equal footing with ordinary periodic points. Indeed, in general, the only "true" dynamics is that taking place in the reduced phase space P_μ.

The following shows that our definition coincides with the standard ones (Smale [4], Robbin [4]).

<u>Theorem</u>. (i) $p \in P$ <u>is a relative equilibrium iff there is a one-parameter subgroup</u> $g(t) \in G$ <u>such that for all</u> $t \in R$, $F_t(p) = \Phi_{g(t)}(p)$ <u>where</u> F_t <u>is the flow of</u> X_H <u>and</u> Φ <u>is the action of</u> G.

(ii) $p \in P$ <u>is a relative periodic point iff there is a</u> $g \in G$, <u>and</u> $\tau > 0$ <u>such that for all</u> $t \in R$, $F_{t+\tau}(p) = \Phi_g(F_t(p))$.

<u>Proof</u>. (i) p is a relative equilibrium iff $\pi_\mu(p)$ is a fixed point for the induced flow on P_μ iff $\pi_\mu(F_t(p)) = \pi_\mu(p)$. If this holds there is a unique curve $g(t) \in G_\mu$ such that $F_t(p) = \Phi_{g(t)}(p)$ since the action of G_μ on $\psi^{-1}(\mu)$ is free. The flow property $F_{t+s}(p) = F_t \circ F_s(p)$ immediately gives $g(t+s) = g(t)g(s)$, so $g(t)$ is a one-parameter subgroup of G_μ. Conversely, if $F_t(p) = \Phi_{g(t)}(p)$ where $g(t)$ is a one-parameter subgroup of G, we must show $g(t) \in G_\mu$. But this follows from invariance of $\psi^{-1}(\mu)$ under F_t and equivariance (see §2 above).

One proves (ii) in a similar way. □

As a result of our definition we have the following theorem of Smale, whose proof has also been simplified by Robbin [4] and Souriau. We present yet another proof.

Theorem. *Let μ be a regular value of ψ. Then $p \in \psi^{-1}(\mu)$ is a relative equilibrium iff p is a critical point of $\psi \times H : P \to \mathcal{G}^* \times \mathbb{R}$.*

Proof. By our definition and the non degeneracy of the symplectic form on P_μ, P is a relative equilibrium iff p is a critical point of $\tilde{H}$, the reduced Hamiltonian. Since we have invariance under G, this is equivalent to p being a critical point of $H|\psi^{-1}(\mu)$ i.e. of $\psi \times H$ (Lagrange multiplier theorem). □

Thus the advantage of passing to P_μ is that relative equilibria really become equilibria and, moreover, we have a Hamiltonian system on P_μ with a (non-degenerate) symplectic form.

In the above theorem, it is necessary that μ be a regular value. For example, in the n-body problem (where $G = SO(3)$), if all the bodies are lined up with velocities headed towards the center of mass, we have a critical point of $\psi \times H$ but the bodies do not travel in circles (theorem 4(i) fails).

There are a number of equivalent ways to rephrase the above result if $P = TM$ and $H = K + V$. (In particular see Smale [4]; some interesting conditions have also been given by O. Lanford.)

Using these ideas, Smale is able to estimate the number of relative equilibria by using Morse theory to count the critical points. The results yield quite interesting information for the n-body problem (see Smale [4], Iacob [2]).

Stability of Relative Equilibria.

Let us recall the classical definition of <u>Liapunov stability</u> (see also lectures 5, 8). Let x be a critical point of a flow F_t; i.e., $F_t(x) = x$. Then x is <u>stable</u> if for every neighborhood U of x there is a neighborhood V of x such that $y \in V$ implies $F_t y \in U$ for all t.

Now we can define stability of relative equilibria as follows:

<u>Definition</u>. Let $p \in P$ be a relative equilibrium of the Hamiltonian vector field X_H. We call p <u>relatively stable</u> if the point p is (Liapunov) stable for the induced flow on the quotient space P/G, (on P/G, p is a fixed point).

<u>Theorem</u>. <u>Let the conditions guaranteeing the symplectic structure on P_μ and the above corollary hold and let $p \in P$ be a relative equilibrium. Let $\tilde{H}$ be the induced Hamiltonian on P_μ. If $d^2\tilde{H}$ is definite at $\pi_\mu(p)$, then p is relatively stable.</u>

<u>Proof</u>. The condition tells us that $\pi_\mu(p)$ is a stable fixed point on P_μ, by conservation of energy. Thus we conclude that within each $\psi^{-1}(\mu)/G_\mu$, p is stable. But by openness of the conditions, the same is true of nearby reduced phase spaces $P_{\mu'}$, μ' near μ. Thus p is actually relatively stable. □

If G is a Lie group with a left invariant metric, a relative equilibrium represents a fixed point v in the Lie algebra, or a one-parameter subgroup of G. We can use theorem 6 to test for its stability. If we do so, we recover a result of V. Arnold [1] (who proved it directly by an apparently more complicated procedure) as follows. The quadratic form $d^2\tilde{H}$ at $v \in \mathcal{G}$ is, in this case, worked out to be -- after a short straightforward computation:

$$Q_v(w_1, w_2) = \langle B(v, w_1), B(v, w_2)\rangle + \langle [w_1, v], B(v, w_2)\rangle$$

where $\langle B(u, v), w\rangle = \langle [u, w], v\rangle$. Thus the condition requires Q_v to be definite. In case of a rigid body ($G = SO(3)$) this yields the classical result that a rigid body spins stably about its longest and shortest principal axes, but unstably about the middle one. For fluids ($G = \mathcal{D}_\mu$ = group of volume-preserving diffeomorphisms) the situation is complicated by the fact that the metric is only weak so the criterion is not directly applicable. In celestial mechanics stability of the relative equilibria often depends on stability criteria much deeper than that above, such as Moser's "twist stability theorem"; (see Abraham [2]).

Completeness of Homogeneous Spaces.

Recall that a homogeneous space is a manifold together with a transitive group action Φ on it. The following is a classical and useful result.

Theorem. Let M be a riemannian manifold and suppose either

(a) M is compact

or (b) M is a homogeneous space, the transitive action consisting of isometries.

        Then M is geodesically complete.

Here, geodesically complete means that the geodesic flow on TM is complete; i.e. geodesics can be indefinitely extended (without running off M). In the finite dimensional case it is equivalent to M being complete as a metric space and to closed balls being compact.

To prove (a) one uses the fact that if an integral curve stays in a compact set then it can be indefinitely extended (this follows from the local existence theory). But TM is a union of compact invariant sets, namely the sets $S^c = \{v \in TM \mid \|v\| = c\}$, $c \in R$, $c \geq 0$. Hence (a) holds.

One proves (b) by using the homogeneity to keep translating vectors to a fixed point say x_0, to estimate the time of existence. This time does not shrink because of conservation of energy. Hence one can keep on extending a geodesic by a definite ε time interval, independent of the base point. Hence a geodesic can be indefinitely extended.

For pseudo-riemannian manifolds (i.e. the metric need not be positive definite) this argument does not work. However we have the following (see Wolf [1], p. 95, Marsden [9]).

Theorem. Let M be a compact pseudo-riemannian manifold. Let G be a Lie group which acts transitively on M by isometries. Then M is geodesically complete.

This result was proved by Hermann [3] in the special case of a semi-simple compact Lie group carrying a left invariant pseudo-riemannian metric. It should be noted that in the statement of the theorem neither the homogeneity nor the compactness may be dropped. For example it has become well-known to relativists that there are incomplete Lorentz metrics on the two torus. These were constructed by Y. Clifton and W. Pohl. (See Markus [1], p. 189.) An incomplete metric on the noncompact group $SO(2, 1)$ is constructed in Hermann [3] although this is a special case of a whole class of incomplete pseudo-riemannian manifolds constructed by J. A. Wolf. (See Wolf [1,2]).

Proof. We shall show that the tangent bundle TM of M is the union of compact subsets S_α parametrized by elements α of the dual $\mathcal{G}^*$ of the Lie algebra of G, with S_α invariant under the geodesic flow. Since a vector field whose integral curves remain in a compact set has a complete flow, this is clearly enough to prove the theorem, as above.

Let $P : TM \to \mathcal{G}^*$, $P(v) \cdot \xi = \langle v, \xi_M(x) \rangle$ be the moment and for $\alpha \in \mathcal{G}^*$, set $S_\alpha = P^{-1}(\alpha)$. By the conservation theorems, S_α is invariant under the flow. Obviously TM is the union of the S_α. Therefore, it remains only to prove the following lemma. In this lemma we use the fact that $T_x M = \{\xi_M(x) | \xi \in \mathcal{G}\}$ which follows from the fact that M is homogeneous, i.e. there is only one orbit.

Lemma. Each of the sets S_α is a compact subset of TM.

Proof. Certainly S_α is closed. Furthermore, the restriction of the canonical projection $\pi : TM \to M$ to S_α is one-to-one because from the fact that the $\xi_M(m)$ span T_mM, we see that S_α intersects each fiber in at most one point.

We claim first of all that $\pi(S_\alpha)$ is closed and hence compact. Indeed $x \notin \pi(S_\alpha)$ means that α is not in the range of the linear map obtained by restricting P to T_xM. Thus α is not in the range of $P|T_yM$ for y in a whole neighborhood of x. Hence $\pi(S_\alpha)$ is closed.

Now let $v_x, v_y \in S_\alpha$, so $\langle v_x, \xi_M(x)\rangle = \langle v_y, \xi_M(y)\rangle = \alpha(\xi)$ for all $\xi \in \mathcal{G}$. From the fact that $\xi_M(m)$ span T_mM and non-degeneracy of $\langle , \rangle$, we may conclude that v_x is close to v_y if x is close to y. Hence the inverse $\pi^{-1} : \pi(S_\alpha) \to S_\alpha$ is continuous. Thus S_α is compact. □

Remarks. 1. If $\dim G = \dim M$, then S_α is actually a submanifold because $P : TM \to \mathcal{G}^*$ is a submersion in that case (the derivative of P along the fibers is one-to-one and hence surjective).

2. Of course we have actually proved more. We only require that the infinitesimal generators span at each point, and that we have an invariant Hamiltonian system. Clearly conservation of energy, which is the basis of the proof for the Riemannian case (see above), plays no role here.

7. Quantum Mechanical Systems.*

In this lecture we shall describe a few aspects of quantum mechanics. Obviously we cannot be exhaustive here, but we will try to mention a number of important foundational points. (For further elucidation, see von Neumann [2], Mackey [1, 2], Jauch [1], Varadarajan [1], Chernoff-Marsden [1]).

In order to clarify the differences between classical and quantum mechanics, it is convenient to adopt a probabilistic point of view and think in terms of statistical mechanics rather than particle mechanics. We begin with some general considerations.

Basic Properties of Physical Systems.

A <u>physical system</u> consists of two collections of objects, denoted S and $\mathcal{O}$ -- called <u>states</u> and <u>observables</u> respectively -- together with a mapping

$$S \times \mathcal{O} \to \text{(Borel probability measures on the real line } R)$$
$$(\psi, A) \mapsto \mu_{A, \psi} .$$

Additionally, there is usually a Hamiltonian structure described below.

Elements $\psi \in S$ describe the state of the system at some instant and elements $A \in \mathcal{O}$ represent "observable quantities"; when A is measured and the system is in state ψ, $\mu_{A,\psi}$ represents the probability distribution for the observed values of A. Thus if $E \subset R$, $\mu_{A,\psi}(E) \in R$ is the probability that we will measure the value of A

---
*This lecture was prepared in collaboration with P. Chernoff.

to lie in the set E if the system is known to be in state ψ.

Normally there is also some dynamics; i.e. a flow or evolution operator $U_t : S \to S$.

The set S is usually a convex set and U_t consists of convex automorphisms. The set P of extreme points -- called the pure states, is usually symplectic and (for conservative systems) the flow F_t on P is Hamiltonian. (The flows F_t, U_t determine one another.)

Statistical Mechanics.

Consider now the following example: let P, ω be a symplectic manifold, say finite dimensional and define the states and observables by:

(a) States; S consists of probability measures ν on P.
(b) Observables; $\mathcal{O}$ consists of real valued functions $A : P \to R$.
(c) The map $S \times \mathcal{O} \to$ (Borel measures on R) is given by
$$\mu_{\nu,A}(E) = \nu(A^{-1}(E)), \text{ where } E \subset R.$$

The states are measures rather than points of P to allow for the fact that we may only have a statistical knowledge of the "exact" state.

It is easy to see that the pure states are point measures, so are in one-to-one correspondence with points of P itself. Note that every observable A is <u>sharp</u> in a pure state; i.e. the corresponding measure on R is a point measure. In other words there is no dispersion when measuring any observable in a pure state.

Around 1930, B. O. Koopman noted that the above picture can be expressed in Hilbert space language. Let $\mathcal{H}$ denote the Hilbert space of all square integrable functions $\psi : P \to C$, with respect to Liouville measure. Each $\psi \in \mathcal{H}$ determines a probability measure $\nu_\psi = |\psi|^2 \mu$ if $\|\psi\| = 1$. If A is an observable, its expected value is

$$\mathcal{E}(A) = \int_P A |\psi|^2 d\mu = \langle A\psi, \psi \rangle$$

where A is regarded as a (self adjoint) multiplication operator on $\mathcal{H}$.

The dynamics $F_t : P \to P$ on phase space P induces in a natural way, and is induced by (under certain conditions) a dynamics on S and on $\mathcal{H}$, namely $U_t \nu = F^*_{-t} \nu$ and $U_t \psi = \psi \circ F_{-t}$.

Consider the map $\psi \mapsto \nu_\psi$ of $\mathcal{H}$ to S . It is many-to-one. In fact $\nu_\psi = \nu_{\psi'}$ if $\psi' = e^{i\alpha} \psi$ where $\alpha : P \to R$. These phase transformations $\psi \mapsto e^{i\alpha} \psi$ form the <u>phase group</u> of classical mechanics.

It is not hard to see that an operator A on $\mathcal{H}$ is a multiplication operator iff it commutes with all phase transformations. Classical observables are those A's which are self-adjoint; i.e. real valued.

Since only the measures have physical meaning, we see that any quantity of physical meaning must be invariant under the phase group.

It follows that the inner products $<\psi, \varphi>$ and their squares $|<\psi, \varphi>|^2$ can have no physical meaning. One says that there is "no coherence" in classical mechanics. (This is because $|<e^{i\alpha}\psi, e^{i\beta}\varphi>| \neq |<\varphi, \psi>|^2$ in general.)

We can think of $\mathcal{H}$ as a symplectic manifold with the usual symplectic form: $\omega = \text{Im}<\ >$. The dynamics induced on $\mathcal{H}$ is unitary and thus symplectic; i.e. it is Hamiltonian (see lecture 2). Thus the dynamics on $\mathcal{H}$ is consistent with the statistical interpretation.

We can regard the phase group G as a symmetry group of $\mathcal{H}$ by $\psi \mapsto e^{i\alpha}\psi$. Indeed, as explained in lecture 6 we can form the reduced phase space; we just get back P and the old dynamics on P. We have the following picture

$$\begin{array}{ccc} \text{Hilbert space} & \mathcal{H} \xrightarrow{\nu_\psi} S & \text{(statistical states)} \\ \text{(Liouville-Koopman picture)} & & \\ & \searrow U & \\ \text{reduction of phase space by the phase group} & P & \text{(pure states)} \end{array}$$

Quantum Mechanics.

Quantum mechanics differs from classical mechanics in that the phase group is much smaller; interference and coherence -- typical wave phenomena -- now play a fundamental role. Furthermore, all predictions are necessarily statistical in that there are no dispersion free states ($\psi \in S$ is <u>dispersion free</u> when $\mu_{A,\psi}$ is a point measure for each $A \in \mathcal{O}$).

In classical mechanics, each state $\nu \in S$ was a "mixture" of pure states. We use ν because of ignorance as to the true state. Increasing our knowledge will "reduce" ν to a measure with smaller variance.

In quantum mechanics, states are not always reducible into statistical states of mixtures. This is clearly illustrated by experiments with polarized beams of coherent light (even with single photons). In such an experiment, states can be described by unit vectors $\psi \in R^2$ giving the direction of polarization. The probability that a φ wave passes through a ψ filter is <u>observed</u> to be $|\langle \varphi, \psi \rangle|^2$. A little thought shows that no such polarized state φ can be realized as a statistical mixture of other polarized states.*

These sorts of experimental facts lead one to consider the states as forming a Hilbert space † $\mathcal{H}$ and the states as being the unit rays in $\mathcal{H}$. (These are the pure states; mixed states corresponding to ν's above are introduced below.) Thus, letting P denote the rays in $\mathcal{H}$ (P is called projective Hilbert space), we have a map $\mathcal{H} \to P$, again many to one. This time the <u>phase group</u> is the circle

---

* Furthermore, the experiment is not reproducible in the sense that no matter how carefully φ is prepared, there is uncertainty in the outcome (unless the probability is 0 or 1). Such an uncertainty seems to be fundamental.

† We take $\mathcal{H}$ to be complex but it is not a priori clear why it shouldn't be real. There are good reasons for the complex structure related to the Hamiltonian structure; (see lecture 2 and references in Jauch [1]).

group $\{e^{i\alpha}; \alpha \in R\}$. The reason P is chosen this way is that one imagines general <u>elementary selective measurements</u> wherein $|\langle \psi, \varphi \rangle|^2$, for each $\psi, \varphi \in \mathcal{H}$, $\|\psi\| = \|\varphi\| = 1$ is the object with physical meaning -- it represents the probability that we will find φ in state ψ or if you like, the "transition probability" for going from φ to ψ.

More generally, we can imagine a general selection measurement. let $F \subset \mathcal{H}$ be a (closed) subspace and $\varphi \in \mathcal{H}$. The probability of transition from φ to F is $\langle P_F \varphi, \varphi \rangle$ where P_F is the orthogonal projection onto F.

Just as in the case of statistical mechanics we observe that P is the reduction of $\mathcal{H}$ by the phase group (this was noted in lecture 6).

Once the above view is accepted, then as Mackey has shown, the rest of the picture of what S, $\mathcal{O}$ and $\mu_{A,\psi}$ have to be is pretty much forced upon us. This goes as follows.

Consider an observable A. For each $E \subset R$ we have $\mu_{A,\psi}(E)$ measuring a probability of observing A to lie in E if the state is ψ. The previous discussion suggests there should be a projection operator P_E^A on $\mathcal{H}$ such that

$$\mu_{A,\psi}(E) = \langle P_E^A \psi, \psi \rangle .$$

Since μ is a probability measure we must have:

$$P^A_\emptyset = 0, \quad P^A_R = I \tag{1}$$

$$\text{and} \quad P^A_{\bigcup_{i=1}^\infty E_i} = \sum_{i=1}^\infty P^A_{E_i} \tag{2}$$

if E_i are disjoint. It follows that the $P^A_{E_i}$ are mutually orthogonal. We also must have by (2),

$$P^A_{E \cup F} = P^A_{E \backslash F} + P^A_{F \backslash E} + P^A_{E \cap F}$$

$$P^A_E = P^A_{E \backslash F} + P^A_{E \cap F}$$

$$\text{and} \quad P^A_F = P^A_{F \backslash E} + P^A_{E \cap F} .$$

Hence $P^A_E P^A_F = P^A_{E \cap F} = P^A_F P^A_E$; i.e. the P^A_E's commute.

The spectral theorem (see, e.g. Yosida [1]) now tells us that there is a unique self adjoint operator, also denoted A, such that $A = \int_{-\infty}^\infty \lambda \, dP^A_\lambda$; $\{P^A_\lambda\}$ is the spectral measure of A. Conversely any self adjoint operator A yields a spectral measure and hence defines $\mu_{A,\psi}$.

Thus, to every observable there is a self adjoint operator A, but it is not clear that every self adjoint operator is physically realizable. (For example it is not clear how to measure (position) plus (momentum) = $q + p$ in the laboratory.)

Of course it is well known that a self adjoint operator (like the position operator) need not have any square integrable eigenfunctions. What is asserted to be of physical relevance is the probability measure $\mu_{A,\psi}$, which is always well defined. Of course, one must avoid trivial "paradoxes" in quantum mechanics which arise from an inadequate understanding of the spectral theorem, or by ascribing more physical meaning (e.g. individual trajectories) to the theory than that given by the $\mu_{A,\psi}$.

Notice that the expected value of A in a state ψ is
$$\mathcal{E}(A) = \int_{-\infty}^{\infty} \lambda \, d\mu_{A,\psi}(\lambda) = \int_{-\infty}^{\infty} \lambda \, d\langle P_\lambda^A \psi, \psi\rangle = \langle A\psi, \psi\rangle.$$

Thus a state φ yields a mapping $F \mapsto \langle P_F \varphi, \varphi\rangle$ of subspaces in $\mathcal{H}$ to $[0, 1]$ describing a transition probability. It is a "probability measure" based on the closed subspaces.

We can generalize the notion of state so as to allow for the possibility of mixed states (with the same statistical interpretation as in the classical case) by just considering a general "measure" defined on the closed subspaces of $\mathcal{H}$. It is a famous theorem of Gleason (see Varadarajan [1] for a proof) that such a state is given by $F \mapsto \text{trace}(P_F D)$ where D is a positive operator of trace one on $\mathcal{H}$, called a <u>density matrix</u>.

Thus quantum mechanics is specified as follows: we are given a complex Hilbert space $\mathcal{H}$ and set

$$\begin{cases} S = \text{all density matrices, a convex set} \\ \mathcal{O} = \text{self adjoint operators on } \mathcal{H} \\ \mu_A(E) = \text{trace } (P_E^A D), \; P_E^A \text{ the spectral projections of } A. \end{cases}$$

It is not hard to see that the pure states (extreme points of S) are identifiable with unit vectors in ℋ, modulo the phase group -- what we previously called P.

Thus we again get this picture:

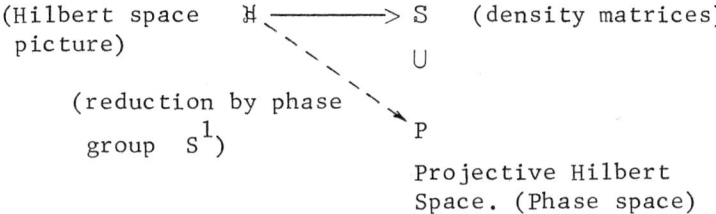

(Hilbert space ℋ ⟶ S (density matrices)
picture)
⋃
(reduction by phase group S^1) ↘ P

Projective Hilbert Space. (Phase space)

Bargmann-Wigner Theorem.

In the case of statistical mechanics we saw that the flow on P naturally induced one on ℋ. For quantum mechanics this is not so obvious, and was considered by Bargmann-Wigner (see Varadarajan [1] for proofs and details). Since P are the extreme points of S, we can just as well work with S as P.

<u>Theorem</u>. <u>Let</u> U_t <u>be a flow by convex automorphisms on</u> S. <u>Then</u> U_t <u>is induced by a one parameter unitary group</u> V_t <u>on</u> ℋ, <u>unique up to phase factors</u>.

The result is conceptually important because ℋ is a mathematical construct for analytical convenience. Only P should be directly physically relevant.

<u>Note</u>. A convex automorphism of S can be implemented by either a unitary or anti-unitary operator (Wigner), but for one parameter groups

the later case is excluded.

One can go on to see when actions of groups lift from P (or S) to $\mathcal{H}$. Then there is a cohomology condition needed on the group (see Simms [1], Chichilnisky [1] for more information).

We also remark that S is, in the classical case, a simplex. This means that each state in S can be uniquely represented in terms of the extreme points (see Choquet [1] for general information on this point).

<u>Miscellany</u>.

At this point one might ask: what do we choose for $\mathcal{H}$? The answer actually is that $\mathcal{H}$ should be L_2 of the configuration (rather than phase) space. The reason for this is best seen through an analysis given by Mackey and Wightman. The result is that if one makes entirely reasonable hypotheses on what the position and momentum operators ought to be, then their structure and that of the Hilbert space is determined. One finds that if the classical phase space is $P = T^*M$, then the quantum mechanical Hilbert space is $\mathcal{H} = L_2(M, \mathbb{C})$ and the quantum operators corresponding to a position observable f (a function $f : M \to \mathbb{R}$) and a momentum observable $P(X)$ ($P(X) \cdot \alpha = \alpha(X)$, X a vector field on M -- see lecture 6) are:

$$Q_f = \text{multiplication by } f$$

and $P_X = iX$ as a differential operator.

The associations $f \mapsto Q_f$, $P(X) \mapsto P_X$ are often called the <u>Dirac</u>

Quantization Rules. They preserve bracket operations.

This suggests that corresponding to the classical system $K + V$ on TM is the quantum system with energy operator $-\Delta + V$ on $\mathcal{H}$. (Some problems related to this are discussed in the next lecture.)

Exact quantum procedures are not so simple. In fact an old theorem of Groenwald and Van Hove asserts that there is no map possible from all classical observables to quantum observables that preserves the bracket operations. However much work is currently being done on some geometric aspects of this problem (see Souriau [1]).

Another fundamental question is the reverse problem: in what sense is classical mechanics a limit of quantum mechanics (as h, Planck's constant $\to 0$) ? This has been investigated by many people, but the deepest analysis seems to be due to Maslov (see Arnold [2]). This problem is discussed further in the Appendix.

The C^* Algebra Approach to Quantum Mechanics.

There are many ways of generalizing the examples of physical systems given in the first part of the lecture. One of these, taken by von Neumann, is to regard the set of observables as an algebra. This is mathematically convenient although it may not correspond exactly with physical reality for as mentioned above, the sum of two observables need not be observable. Other ways of generalization are the "quantum logic" point of view described in Varadarajan [1] and Mackey [1].

In the classical case the algebra is the algebra of functions on phase space -- a commutative algebra. The quantum case is distinguished by having a non-commutative algebra. Indeed any C* algebra which is commutative must be isomorphic to a space of continuous functions and so, is in this sense, classical.

Segal's formulation of this point of view proceeds as follows. Let $\mathcal{A}$ be a C* algebra; i.e. a Banach space which is also an algebra and has a conjugation (or adjoint) operation * satisfying certain simple axioms. For example one can think of an algebra of bounded operators on a Hilbert space (unbounded operators are included via their spectral projections). Simmons [1] contains a very readable account of the elementary properties of C* algebras.

Take the <u>observables</u> to be the self adjoint elements of $\mathcal{A}$.

The <u>states</u> are the normalized positive linear functionals on $\mathcal{A}$. (It is easy to see that they are automatically continuous.) We are to think of states in the same way as before. If $\mathcal{E}$ is a state, $\mathcal{E}(A)$ is the expectation of A in the state $\mathcal{E}$.

Of central importance is the Gelfand-Naimark-Segal construction: Let $\mathcal{A}$ be a C* algebra and $\mathcal{E}$ a state of $\mathcal{A}$. Then there is a Hilbert space $\mathcal{H}$, a unit (cyclic) vector $\psi \in \mathcal{H}$ and a *-representation $\pi_\mathcal{E} : \mathcal{A} \to \mathcal{L}(\mathcal{H})$ (the bounded operators on $\mathcal{H}$) such that

$$\mathcal{E}(A) = \langle \pi_\mathcal{E}(A)\psi, \psi \rangle \quad \text{for all} \quad A \in \mathcal{A}.$$

In fact $\mathcal{H}$, ψ, π are unique up to unitary equivalence. See Lanford [1] for details.

In this way, we can construct our probability measure $\mu_{A,\mathcal{E}}$. Thus we have a general example of a physical system consisting of S, $\mathcal{O}$ and the map $\mu_{A,\mathcal{E}}$ just constructed which includes both classical and quantum systems as special cases.

There is no canonical Hilbert space, but one can be constructed for each $\mathcal{E}$. We can still form P, the extreme points of S, but in general P won't be a symplectic manifold. (It is in the examples previously constructed however.)

The above Gelfand-Naimark-Segal construction is similar to Gleason's theorem in that it delineates states. It essentially enables one to recover the Hilbert space formalism from the abstract C* algebra formalism. However, often it is convenient to stick with the general C* algebra point of view. For example, one can characterize pure states $\mathcal{E}$ as those for which $\pi_\mathcal{E}$ is irreducible.

Several other ideas from the Hilbert space approach carry over. For example the general form of the uncertainty principle is valid: for observables A, $B \in \mathcal{Q}$, and a state $\mathcal{E}$,

$$\sigma(A, \mathcal{E})\sigma(B, \mathcal{E}) \geq \tfrac{1}{2}\mathcal{E}(C) \,, \quad C = i(AB - BA)$$

where $\sigma(A, \mathcal{E})$ is the variance of the probability distribution $\mu_{A,\mathcal{E}}$: $\sigma(A, \mathcal{E})^2 = \mathcal{E}(A^2) - (\mathcal{E}(A))^2 = \mathcal{E}((A - \mathcal{E}(A)I)^2)$.

Proof. Let $[X, Y] = \mathcal{E}(XY^*)$. This is an inner product on $\mathcal{Q}$ so obeys the Schwartz inequality. Note that it is enough to prove the inequality in case $\mathcal{E}(A) = 0$, $\mathcal{E}(B) = 0$ for we can replace A, B by A - $\mathcal{E}(A)$I, and B - $\mathcal{E}(B)$I. Then

$$\mathcal{E}(C) = i[\mathcal{E}(AB) - \mathcal{E}(BA)]$$
$$= 2 \text{ Im } [A, B]$$
$$\leq 2 [A, A]^{\frac{1}{2}}[B, B]^{\frac{1}{2}}$$

so $\frac{1}{2}\mathcal{E}(C) \leq \sigma(A, \mathcal{E})\sigma(B, \mathcal{E})$, Q E D.

A Hidden Variables Theorem.

The orthodox interpretation of quantum mechanics presented above has discomforted many physicists, notably including Planck, Einstein, de Broglie, and Schrodinger (see for example De Broglie [1] and Einstein-Podolsky-Rosen [1]). It is hard to escape the feeling that a statistical theory must be, in some sense, an incomplete description of reality. One might hope that the probabilistic aspects of the theory are really due, as in the case of classical statistical mechanics, to some sort of averaging over an enormous number of "hidden variables"; in a perfect description of a state, in which these hidden parameters would have well-determined values, all the observables would be sharp. However, von Neumann [2] has given a proof that the results of quantum mechanics are not compatible with a reasonably formulated hidden variable hypothesis. We shall outline an argument along von Neumann's lines, but in the more general setting of Segal's C*-algebra formulation of quantum theory.

Let the observables of a given physical system be represented by the self-adjoint elements of a C* algebra $\mathcal{A}$. If $A \in \mathcal{A}$ is an observable and ρ is a state, the dispersion of A in the state ρ is given by $\sigma^2(A, \rho) = \rho(A^2) - \rho((A - \rho(A)I)^2)$. We shall say that ρ is a <u>dispersion-free state</u> provided that $\sigma^2(A, \rho) = 0$ for every observable $A \in \mathcal{C}$. The results of experiment show that the states of quantum systems prepared in the laboratory are not dispersion-free. The hidden-variable hypothesis is that the physical state ρ owes its dispersion to the fact that it is a statistical ensemble of ideal dispersion-free states. (The latter need not be physically realizable -- just as one cannot really prepare a classical gas with precisely determined positions and velocities for each of its molecules.) Mathematically, the hypothesis states that every state ρ is of the form

(1) $$\rho(A) = \int_\Omega \rho_\omega(A) \, d\mu(\omega)$$

where each ρ_ω is a dispersion-free state and μ is a probability measure on some space Ω. The coordinate $\omega \in \Omega$ represents, of course, the indeterminate "hidden variables".

<u>Theorem.</u> (See Segal [4].) <u>A C* algebra $\mathcal{C}$ admits hidden variables in the above sense only if $\mathcal{C}$ is abelian.</u> (The corresponding physical system is then "classical".)

<u>Proof.</u> (Chernoff) The first step is to show that a dispersion-free state ρ_ω is multiplicative. Note that the bilinear form $\langle\langle A, B \rangle\rangle = \rho_\omega(AB^*)$ is a Hermitian inner product on $\mathcal{C}$. ($\langle\langle A, A \rangle\rangle = \rho_\omega(AA^*)$ is ≥ 0 by hypothesis. From this it follows

easily that $\rho_\omega(C^*) = \overline{\rho_\omega(C)}$ for any $C \in \mathcal{A}$. In particular we have $\langle\langle B, A\rangle\rangle = \rho_\omega((AB^*)^*) = \overline{\rho_\omega(AB^*)} = \overline{\langle\langle A, B\rangle\rangle}$.) Hence, by the Schwarz inequality,

$$|\rho_\omega(AB)| \leq \rho_\omega(AA^*)^{\frac{1}{2}} \rho_\omega(B^*B)^{\frac{1}{2}}$$

for all $A, B \in \mathcal{A}$. From this we see that if $\rho_\omega(AA^*) = 0$ then $\rho_\omega(AB) = 0$ for all B. Suppose that A is self-adjoint. Then, since $\mathcal{A}$ is dispersion-free, $\rho_\omega((A - \rho_\omega(A)I)^2) = 0$. Therefore, for every B, $\rho_\omega((A - \rho_\omega(A))B) = 0$. That is, $\rho_\omega(AB) = \rho_\omega(A)\rho_\omega(B)$. This holds as well for non-self-adjoint A by linearity. In particular, if $\mathcal{A}$ is dispersion-free it follows that $\rho_\omega(AB) = \rho_\omega(BA)$.

But if $\mathcal{A}$ admits hidden variables, it follows immediately from (1) that every state ρ satisfies $\rho(AB) = \rho(BA)$. Since there are enough states to distinguish the members of $\mathcal{A}$, (e.g. states of the form $A \mapsto \langle A\psi, \psi\rangle$) it follows that $AB = BA$. Thus $\mathcal{A}$ is abelian. □

Remark. Conversely, a well-known theorem of Gelfand and Naimark states that every abelian C* algebra is isomorphic to $C(X)$, the set of continuous functions on some compact set X. (Many accounts of this result are available; a very readable one is in Simmons [1].) The states of $\mathcal{A}$ are simply the probability measures on X, which are convex superpositions of the δ-measures at the points of X; the latter are, of course, precisely the dispersion-free states.

We can also dispose of a less stringent notion of hidden variables. According to Jauch [11], Mackey has proposed the consideration of "ε-dispersion-free" states. A state ρ is called ε-dispersion-free if for every <u>projection</u> $E \in \mathcal{A}$ we have $\rho^2(E, \rho) < \varepsilon$. A system is said to admit "quasi-hidden variables" if for all $\varepsilon > 0$, every state can be represented as $\int \rho_\omega d\mu(\omega)$ where all the states ρ_ω are ε-dispersion-free. (In fact one can say "if for some $\varepsilon > 0$ sufficiently small" but this leads to a harder theorem). If $\mathcal{A}$ admits quasi-hidden variables and ρ is a <u>pure state</u> of $\mathcal{A}$, then it is easy to see that ρ is ε-dispersion free for every ε. Then by the argument above ρ must be multiplicative on the algebra generated by the projections in $\mathcal{A}$. This will be all of $\mathcal{A}$ in many interesting cases -- in particular, if $\mathcal{A}$ is a von Neumann algebra (i.e. closed in the strong operation topology). But then, because the pure states separate elements of $\mathcal{A}$, it follows as before that $\mathcal{A}$ is abelian. (We must hasten to add that Jauch and Mackey were considering these questions in the context of lattices of "questions" which are more general than the projection lattices which we have discussed; so from the foundational point of view the notion of quasi-hidden variables has raised problems which our simple argument cannot handle.)

The essential point of the argument given above was the non-existence in general of a large supply of linear functionals on $\mathcal{A}$ which carry squares to squares. A much deeper analysis has been carried out by Kochen and Specker [11], cf. also Bell [1]. They have faced squarely the fact, which we have mentioned, that it is really not physically reasonable for the sum of non-commuting observables always to be an observable. Drastically reducing the algebraic operations which they allow, they nevertheless reach the same results; their functionals are required to be linear only on <u>commuting</u> observables. We shall not go into the details of their arguments, for which we refer the reader to their paper, which also includes an interesting discussion of the entire problem of hidden variables and various attempts to introduce them. Some recent work on this subject centering around Bells' inequality has been done. The results again are against hidden variables theories. See Clauser et al [1] and Freedman-Clauser [1].

<u>The Measurement Process</u>.

Let us now discuss the process of measurement in some detail, following von Neumann [1]. (A clear summary of von Neumann's ideas may be found in the book of Nelson [2]; see also Jauch [1] and de Broglie [1].)

Various solutions of the problems of measurement have been proposed; cf. Bohm and Bub [1]. However it is not yet clear that the problems have been solved. The measurement of an observable involves the interaction of a "physical system" with an "observing apparatus", so we should first describe the mathematical treatment of such <u>composite systems</u>.

If the pure states of a system S correspond to the unit rays of $\mathcal{H}$, and those of a second system S' correspond to the rays of $\mathcal{H}'$, then the pure states of the compound system consisting of S and S' correspond to the unit rays of the <u>tensor product</u>* $\mathcal{H} \otimes \mathcal{H}'$.

---

* The tensor product $\mathcal{H} \otimes \mathcal{H}'$ is the direct product in the category of Hilbert spaces, just as the cartesian product is in the category of manifolds (if P and P' are phase spaces for isolated systems $P \times P'$ is the phase space for the interacting system). A pure state in a composite quantum system is much more complicated than an ordered pair of pure states of the subsystems. This fact seems related to many, if not all, of the so-called "paradoxes" of quantum theory.

The tensor product of Hilbert spaces $\mathcal{H}$ and $\mathcal{H}'$ is by definition the completion of their algebraic tensor product with respect to the following inner product:

$$\langle \sum_i \varphi_i \otimes \varphi'_i, \sum_j \psi_j \otimes \psi'_j \rangle = \sum_{i,j} \langle \varphi_i, \psi_j \rangle \langle \varphi'_i, \psi'_j \rangle .$$

For example, $L^2(R^3) \otimes L^2(R^3) = L^2(R^6)$. If $\{e_i\}$ and $\{f_j\}$ are orthonormal bases of $\mathcal{H}$ and $\mathcal{H}'$ respectively, then $\{e_i \otimes f_j\}_{i,j=1}^{\infty}$ is an orthonormal basis of $\mathcal{H} \otimes \mathcal{H}'$. An observable A of S corresponds to the operator $A \otimes I$ on $\mathcal{H} \otimes \mathcal{H}'$; similarly the observable B of S' corresponds to $I \otimes B$. It can be shown that every observable of the composite system is a function of observables of the above sort, in the sense that every bounded operator on $\mathcal{H} \otimes \mathcal{H}'$ is a limit of operators of the form $\sum (A_i \otimes I) \cdot (I \otimes B_i)$. A state ρ of the compound system determines a state of S by the relation

$$\rho_S(A) = \rho(A \otimes I) .$$

It is important to note that ρ_S will in general be a mixture even if ρ is pure. Thus, if ρ is given by the vector $\sum \varphi_1 \otimes \varphi'_i$, with $\{\varphi_i\}$, $\{\varphi'_i\}$ orthogonal systems in $\mathcal{H}$ and $\mathcal{H}'$, we have

$$\rho_S(A) = \sum \|\varphi'_i\|^2 \langle A\varphi_i, \varphi_i \rangle ,$$

so that ρ_S is given by the density matrix $\sum \|\varphi'_i\|^2 P_{\varphi_i}$.

Now let S be a physical system which we wish to study.

Suppose that we wish to measure an observable A of S. For simplicity let us assume that A has a pure point spectrum, with eigenvectors $\varphi_1, \varphi_2, \ldots$. To measure A it is necessary to allow the system S to interact with an apparatus S'. A suitable apparatus for measuring A will have the property that, if the system S is initially in the state φ_i, after the interaction the composite system of S and S' will be in the state $\varphi_i \otimes \theta_i$, where $\{\theta_i\}$ is a sequence of orthonormal vectors in $\mathcal{H}'$. The interaction, of course, is governed by the Schrodinger equation for the composite system. Hence, if the initial state of S is given by $\psi = \sum_1^\infty c_i \varphi_i$, the final state of $S + S'$ will be $\theta = \sum_1^\infty c_i \varphi_i \otimes \theta_i$ by linearity. Now if B is an observable of S', then after the interaction the expected value of B will be

$$\langle (I \otimes B)\theta, \theta \rangle = \sum_{i=1}^{\infty} |c_i|^2 \langle B\theta_i, \theta_i \rangle ,$$

so that, although $S + S'$ is in the pure state θ, S' is in the mixed state $\sum_{i=1}^{\infty} |c_i|^2 P_{\theta_i}$. Similarly, S is in the mixed state $\sum_{i=1}^{\infty} |c_i|^2 P_{\varphi_i}$.

Now the apparatus is supposed to be of a macroscopic nature, its orthogonal states θ_i represent, say, different counter readings. After the interaction the observer "looks" at the apparatus. Through his faculty of introspection he realizes that the apparatus is in a definite state, say θ_j. (This occurs with probability $|c_j|^2$.)

Once this act of consciousness has taken place it is no longer true that the state of $S + S'$ is $\sum_{i=1}^{\infty} c_i \varphi_i \times \theta_i$; it must be $\varphi_j \times \theta_j$. One then says that the system has been found to be in the state φ_j. This is the famous (or notorious) "reduction of the wave packet".*

We now venture to make some philosophical remarks. It is important to realize that an analogous "reduction" takes place in a <u>classical</u> statistical mechanical system when new information is gained. This is never regarded as a difficulty, because the classical probability packet is always viewed as a mere reflection of the observer's ignorance of the objective underlying state of the system. This is a perfectly consistent interpretation. Why can not the same interpretation serve in the quantum mechanical case?

As long as we are concerned only with a <u>single observable</u> (or with **a** commuting family of observables) it is perfectly possible to view the quantum system classically. That is, one can interpret the reduction from the mixture to the state φ_j as a reduction of classical type. But the existence of incompatible observables in quantum mechanics forces this interpretation to break down. Indeed, the entire point of the negative results concerning "hidden variables"

---

* Of course, "looking at the apparatus" involves interaction with some further apparatus ultimately with the consciousness of the observer. But one can lump all that into S and the observers mind into S'. Nevertheless, apparently one cannot find a mathematical device to yield the reduction of pure states. This is the fundamental problem in interpreting the foundations of quantum mechanics.

is that there is <u>no</u> "objective underlying state" of the system!

Perhaps the quantum probability distributions can be interpreted as reflecting our partial knowledge, as long as we do not insist that there be an objective entity of which we have partial knowledge. This seems reminiscent of the problem of the golden mountain in the sentence "The golden mountain does not exist". If one asks "what does not exist?" and answers "the golden mountain", one is implying that the golden mountain is in fact an entity with some sort of "existence". Some philosophers tried to rescue the situation by stating that the golden mountain "subsists" -- that is, has enough of a shadowy sort of existence to serve as the subject of a sentence. Now Bertrand Russell has observed that the real solution of the problem is to recognize that the original sentence is implicitly quantified, and actually should be regarded as saying "for every x it is false that x is both golden and mountainous". In the absence of new physical discoveries, it seems not impossible that the same sort of purely grammatical trick may be the ultimate solution of the quantum measurement problem.

8. Completeness Theorems and Nonlinear Wave Equations.

As we have mentioned in earlier lectures, the question of the completeness of a flow is a fundamental one; i.e. can solutions be indefinitely extended in time?

In order to deal with this question, one usually proceeds as follows. One first establishes a local existence theorem and then one uses some kind of estimates (so called a priori estimates) to show that this solution does not move to ∞ in a finite time, and hence can be extended to exist for all time. (See the results at the end of lecture 6.)

Below we shall illustrate this general procedure with a couple of examples. We begin by describing a general technique based on energy estimates.

Liapunov Methods.

The concept of Liapunov stability (see lecture 6) can be used effectively as a completeness theorem. Below we shall apply this theorem to nonlinear wave equations.

<u>Theorem</u>. <u>Let E be a Banach space and F_t a local flow on E with fixed point at 0. Suppose that for any bounded set $B \subset E$ there is an $\epsilon > 0$ such that integral curves beginning in B exist for a time interval $\geq \epsilon$.</u>

<u>Let $H : E \to R$ be a smooth function invariant under the flow.</u>
(a) <u>If $H(u) \geq \text{const.} \|u\|^2$, then the flow is complete.</u>

(b) __If__ $H(0) = 0$, $DH(0) = 0$ __and__ $D^2H(0)$ __is positive or negative definite, then there is a neighborhood__ U __of__ 0 __such that any integral curve starting in__ 0 __is defined for all__ t; __moreover,__ 0 __is stable.__

Proof. (a) Let $u \in E$. Since H is conserved we have the a priori estimate $\|u\|^2 \leq$ constant, so u remains in a Bounded set B. But because of the assumption on the flow, the integral curve beginning at u can be indefinitely extended.

(b) From the assumptions, there are constants α, β such that

$$\alpha \|u\|^2 \leq |D^2 H(0) \cdot (u, u)| \leq \beta \|u\|^2.$$

Hence, by Taylor's theorem, in a small neighborhood U_0 of 0, we have

$$\gamma \|u\|^2 \leq |H(u)| \leq \delta \|u\|^2.$$

Because H is conserved, this shows that there are neighborhoods U, V of 0 such that if $u \in U$, it remains in V as long as it is defined. Hence we have completeness as in (a). Since V can be arbitrarily small, we also have stability. □

Nonlinear Wave Equations.

The following equation has been of considerable interest in quantum field theory:

$$\frac{\partial^2 \varphi}{\partial t^2} = \nabla^2 \varphi - m^2 \varphi - \alpha \varphi^p \tag{1}$$

on R^n, where φ is a scalar function, $m > 0$, $\alpha \in R$ and $p \geq 2$ is an integer. The constant α is called the coupling constant and the non-linear term $\alpha\varphi^p$ represents some sort of self interaction of the field φ.

This equation in the same sense as the linear wave equation (see lecture 2) is Hamiltonian. The energy function is

$$H(\varphi, \dot\varphi) = \tfrac{1}{2}\langle\dot\varphi, \dot\varphi\rangle + \tfrac{1}{2}\langle\nabla\varphi, \nabla\varphi\rangle + \frac{m^2}{2}\langle\varphi, \varphi\rangle + \frac{\alpha}{p+1}\int \varphi^{p+1} dx .$$

We chose the phase space to be $H^1 \times L_2$ as for the linear wave equation.

We want to apply the previous theorem to discuss global solutions. In order to do this we need a local existence theory and we need to know H is smooth. For the latter, the key thing is whether or not φ^{p+1} is integrable. To answer this one uses a generalization of the Sobolev inequalities. We shall discuss these points in turn, but let us first state the results corresponding to cases (a), (b) of the previous theorem.

<u>Theorem</u>. (a) <u>Suppose</u> $n = 2$, $\alpha > 0$ <u>and</u> p <u>is odd, or else</u> $n = 3$ <u>and</u> $p = 3$. <u>Then the flow of</u> (1) <u>is complete</u>.

(b) <u>Suppose</u> $n = 2$ <u>with</u> p, α <u>arbitrary or</u> $n = 3$, $p = 2, 3, 4$, α <u>arbitrary. Then there is an</u> $\varepsilon > 0$ <u>such that if</u> $\varphi, \dot\varphi$ <u>is in the</u> $H^1 \times L_2$ ε<u>-ball about</u> 0 <u>then the corresponding solutions exist for all</u> $t \in R$ (<u>actually if the initial data is</u> C^∞, <u>so is the solution</u>). <u>Furthermore the</u> 0 <u>solution is Liapunov stable in the</u> $H^1 \times L_2$ <u>topology</u>.

Notice that the conditions p odd, $\alpha > 0$ is precisely what makes the last term of $H \geq 0$, so $H(\varphi, \dot{\varphi}) \geq \text{const}(\|\varphi\|^2_{H^1} + \|\dot{\varphi}\|^2_{L_2})$ which is (a) of the previous theorem.

The other restrictions on n, p come from the Sobolev theorem in the following form. (See Nirenberg [1], Cantor [1]).

Sobolev-Nirenberg-Gagliardo inequality: Suppose

$$\frac{1}{p} = \frac{j}{n} + a(\frac{1}{r} - \frac{m}{n}) + (1 - a)\frac{1}{q}$$

where $\frac{j}{m} \leq a \leq 1$ (if $m - j - \frac{n}{r}$ is an integer ≥ 1, only $a < 1$ is allowed). Then for $f : R^n \to R^k$,

$$\|D^j f\|_{L^p} \leq (\text{const}) \|D^m f\|^a_{L^r} \cdot \|f\|^{1-a}_{L^q}$$

for a constant independent of f.

For example suppose $n = 3$ and $f \in H^1$. Then taking $m = 1$, $r = 2$, $q = 2$, $j = 0$, $a = 1$ we find that $f \in L^6$ and

$$(\int f^6 dx)^{1/6} \leq (\text{const})(\int (\nabla f)^2)^{1/2} .$$

Such results can be used to prove smoothness of H above and smoothness results in the following:

Local Existence Theory.

Theorem. Let E be a Banach space, $A : D \subset E \to E$ linear, the generator of a semi-group U_t and let $J : E \to E$ be smooth with DJ

bounded on bounded sets. Then

$$\frac{du}{dt} = Au + J(u)$$

defines a unique local flow whose local time of existence is uniformly > 0 on bounded sets. (The evolution operator, F_t is C^∞ for fixed t.)

This result is due to Segal [1] who, based on earlier work of Jorgens, pointed out how it can be used to prove the results (a) on the wave equation (the result (b) is due to, amongst others, Chadam [1], Marsden [10]).

The proof of this result is remarkably simple. Namely, we convert the differential equation to the following integral equation:

$$u(t) = U_t u_0 + \int_0^t U_{t-s} J(u(s)) ds . \qquad (2)$$

The key thing is that the unbounded operator A now disappears and only the bounded operator U_t and the smooth operator J are involved. One can now use the usual Picard method to solve (2). Also one verifies that the solution lies in D if u_0 does and that the solution satisfies the equation (for the latter, J should be C^1 and not merely Lipschitz).

The point is that using the Sobolev-Nirenberg-Gagliardo inequalities one can verify that J has the requisite smoothness: take

$$u = \begin{pmatrix} \varphi \\ \dot\varphi \end{pmatrix} \in H^1 \times L_2$$

$$Au = \begin{pmatrix} -m^2 I & I \\ \Delta & 0 \end{pmatrix}$$

$$Ju = \begin{pmatrix} 0 \\ -\alpha\varphi^p \end{pmatrix}$$

(so one has to check $\varphi \mapsto \varphi^p$ of H^1 to L_2 is smooth). Then the global existence claims follow by the Liapunov method.

We hasten to add that the method depends crucially on the positivity of the linearized energy norm. For other systems of interest, such as the coupled Maxwell-Dirac equations these ideas can give local solutions but they do not help determine if one has global solutions. That particular problem remains largely open. (See Gross [1].)

Quantum Mechanical Completeness Theorems.

Recall that Stone's theorem asserts that every self adjoint operator H on a Hilbert space $\mathcal{H}$ determines a one parameter unitary group (or flow) $U_t = e^{itH}$, defined for all $t \in R$. "Completeness" therefore amounts to the question of verifying self adjointness. Actually this is not such a simple question and is an active area of current research. (See, e.g. Simon [1].)

Let us recall a couple of definitions. Let $\mathcal{H}$ be a Hilbert space and $H : D \subset \mathcal{H} \to \mathcal{H}$ a linear operator, with D dense.

The adjoint $H^* : D^* \subset \mathcal{H} \to \mathcal{H}$ is defined as follows:

$$D^* = \{x \in \mathcal{H} \mid \exists z \in \mathcal{H} \text{ such that } \langle z,y \rangle = \langle x, Hy \rangle \text{ for all } y \in D\}$$

and $H^*x = z$.

An operator is <u>symmetric</u> if $\langle Hx, y\rangle = \langle x, Hy\rangle$ for all $x, y \in D$. Equivalently, $H^* \supset H$; i.e. $D^* \supset D$ and $H^* = H$ on D.

An operator is <u>self adjoint</u> if $H^* = H$.

Often self adjointness is not so easy to check because it depends crucially on the correct choice of D. For example Δ is self adjoint on $H^2(R^n)$, but not on C_0^∞ ($\mathcal{H} = L_2(R^n)$.)*

One is led to introduce another notion. Recall that the <u>closure</u> $\bar{H}$ of an operator H is that operator whose graph is the closure of the graph of H. (This operator $\bar{H}$ always is well defined for symmetric operators.)

A symmetric operator H is called <u>essentially self adjoint</u> if its closure $\bar{H}$ is self adjoint.

It can be shown that this is equivalent to saying that H has <u>at most one self adjoint extension</u>.

For example, Δ with domain C_0^∞ is essentially self adjoint and its closure is Δ with domain H^2.

Since there is a unique way of recovering a self adjoint operator from an essentially self adjoint one, there is no loss in trying to verify the condition of essential self adjointness. This is what is done in practice.

If an operator is not essentially self adjoint this means some

---

*$C_0^\infty = C^\infty$ functions with compact support.

additional information e.g. boundary conditions -- must be specified in order to uniquely determine the dynamics.

Consider the Hamiltonian operator for the Hydrogen atom, $H = -\Delta + \frac{1}{r}$ on R^3. Despite the fact that the solution of the Schrodinger equation has been known explicitly for a half century, only in 1950 was this operator shown to be essentially self adjoint on C_0^∞ (the domain of the closure turns out to be H^2). This was done by T. Kato (see Kato [6] for details and references).

More generally, on R^3, $-\Delta + V$ is essentially self adjoint if $V \in L_2 + L_\infty$. The L_∞ part is trivial, being a bounded operator. To handle the L_2 part (the part of $1/r$ near the origin) one uses a Sobolev estimate in the form: let $V \in L_2$. Then for all $\varepsilon > 0$ there is an M_ε such that for all $f \in H^2$,

$$\|Vf\|_{L_2} < M_\varepsilon \|f\|_{L_2} + \varepsilon \|\Delta f\|_{L_2} .$$

One can then use:

<u>Kato's Criterion.</u> <u>Let</u> A <u>be (essentially) self adjoint on</u> $\mathcal{H}$ <u>with</u> <u>domain</u> D_A. <u>Let</u> B <u>be symmetric,</u> $D_B \supset D_A$ <u>and assume for some</u> $0 < \lambda < 1$

$$\|Bx\| \leq C\|x\| + \lambda\|Ax\|$$

<u>for all</u> $x \in D_A$. <u>Then</u> $A + B$ <u>is (essentially) self adjoint on</u> D_A.

This result is a rather elementary result in operator theory.

We won't go into the details here.

The above method is the basic one by which one handles local singularities such as occur in the Hydrogen atom. On the other hand there can be problems at ∞ such as occur when an atom is placed in an external field. This situation is covered by a theorem of Ikebe-Kato [1]:

<u>Theorem</u>. <u>Let</u> $V : R^3 \to R$ <u>be such that</u> V <u>is smooth and</u> $V(x) \geq V_0(\|x\|)$ <u>where</u> $V_0(r)$ <u>is monotone decreasing and for</u> $H > V_0$,

$$\int^\infty \frac{dr}{\sqrt{H - V(r)}} = +\infty .$$

<u>Then</u> $-\Delta + V$ <u>is essentially self adjoint on</u> C_0^∞ (<u>the</u> C^∞ <u>functions with compact support</u>).

<u>Note</u>. If $V_0(r) = -r^\alpha$, $\alpha \leq 2$ then we have the validity of the assumptions.

The result is too intricate to go into here (an exposition of the proof, generalized to manifolds will appear in Chernoff-Marsden [1]).

Ikebe-Kato then go on to combine this result with the previous type of result. The final result covers most (non-relativistic) cases of interest.

<u>A Classical Analogue of the Ikebe-Kato Theorem</u>.

There is a theorem in classical mechanics which yields

completeness of a Hamiltonian system under the same conditions as in the Ikebe-Kato theorem. (See Weinstein-Marsden [1].) The argument works well on manifolds just as easily as on R^n.

Let us begin by considering the one dimensional case.

Let R^+ be the nonnegative reals and $V_0 : R^+ \to R$ a nonincreasing C^1 function. Consider the Hamiltonian system with the usual kinetic energy and potential V_0; i.e. if $c(t)$ is a solution curve we have

$$c''(t) = -\frac{dV_0}{dx}(c(t)).$$

By monotonicity of V_0, if $c'(0) \geq 0$ then $c'(t) \geq 0$ for all $t \geq 0$. Thus if $H = [c'(t)/2]^2 + V_0(c(t))$ is the constant total energy,

$$c'(t) = 2(H - V_0(c(t)))^{\frac{1}{2}}.$$

<u>Definition</u>. The potential V_0 is <u>positively complete</u> iff

$$\int_{x_1}^{x} \frac{dx}{(2(H - V_0(x)))^{1/2}} \to \infty \text{ as } x \to \infty$$

for all $x_1 \geq 0$ and H such that $V_0(x_1) < H$.

It is easy to see that if this holds for some x_1, H, such that $V_0(x_1) < H$ then it holds for all such x_1, H (use the fact that improper integrals with asymptotic integrands are simultaneously convergent or divergent).

Since the above integral is just the time required for $c(t)$ to move from x_1 to x we see that V_0 is positively complete iff all integral curves $c(t)$ with $c(0) \geq 0$, $c'(0) \geq 0$ are defined for all $t \geq 0$. (The case when $c'(0) = 0$ is easily disposed of.)

Below we will use the notation $\tilde{c}(x_0, H)(t)$ for the integral curve with $\tilde{c}(x_0, H)(0) = x_0$ and energy H.

Example. The function $-x^\alpha$ for $\alpha \geq 0$ is positively complete iff $\alpha \leq 2$. The same is true for

$$-x[\log(x+1)]^\alpha, \quad -x\log(x+1)[\log(\log(x+1)+1]^\alpha \quad \text{etc.}$$

Consider now the general case.

Theorem.* Let M be a complete Riemannian manifold (actually M may be infinite-dimensional) and let V be a C^1 function on M. Suppose there is a point $p \in M$ and a positively complete V_0 on R^+ such that for all $m \in M_1$ [with $d(m, p)$ sufficiently large], $V(m) \geq V_0(d(m, p))$ where d is the Riemannian distance on M. Then the flow on TM of the Hamiltonian vector field with (the usual kinetic energy $K(v) = \langle v, v \rangle/2$ and) potential V is a complete flow (i.e. integral curves are defined for all $t \in R$).

Examples. If $V(m) \geq -(\text{Constant})d(m, p)^2$ for sufficiently large $d(m, p)$ the conditions hold. This is satisfied if $\|\text{grad } V(m)\| \leq (\text{Constant})d(m, p)$ (for sufficiently large $d(m, p)$).

Proof of Theorem. Let $c : [0, b[\to TM$ be an integral curve, $0 < b < \infty$.

---
*See W. Gordon [1], D. Ebin [2], and A. Weinstein and J. Marsden [1].

As usual, it will suffice to show that the curve $c_0(t)$, the projection of $c(t)$ on M, remains in a bounded set for all $t \in [0, b[$ (a similar argument holds for $t \in]-b, 0]$). (In infinite dimensions one uses an argument of Ebin [2].)

Let $n = c_0(0)$ and H the energy of $c(t)$. Let

$$f_1(t) = d(c_0(t), p) \quad \text{and} \quad f_2(t) = \tilde{c}(d(n, p), H)(t)$$

(notation as above). Now

$$f_1(t) \leq d(p, n) + \int_0^t \|c(s)\| ds$$

$$= d(p, n) + \int_0^t (2[H - V(c_0(s))])^{1/2} ds$$

$$\leq d(p, n) + \int_0^t (2[H - V_0(f_1(s))])^{1/2} ds .$$

Also we have

$$f_2(t) = d(p, n) + \int_0^t (2[H - V_0(f_2(s))])^{1/2} ds .$$

It follows from these and monotonicity of V_0 that

$$f_1(t) \leq f_2(t) \leq \tilde{c}(d(n, p), H)(b)$$

for all $t \in [0, b[$. This is an elementary comparison argument. (See the lemma below.) We conclude that $f_1(t) = d(c_0(t), p)$ remains bounded for $t \in [0, b[$ and so the result follows. □

Remarks. (1) The completeness for $t \geq 0$ is preserved if a dissipative vector field Y is added to the Hamiltonian vector field (i.e. Y is vertical $[T\pi(Y) = 0]$ and $Y \cdot K \leq 0$ where K is the kinetic energy). This is easy to see.

(2) This proof also gives an estimate for the growth of $d(c_0(t), p)$ in terms of V_0; for example if $V_0 = -x^2$ then $d(c_0(t), p) - d(n, p)$ grows like e^t.

When is the Sum of two Complete Vector Fields Complete?

Unfortunately, not always. For example consider

$$X : R^2 \to R^2, \quad Y : R^2 \to R^2,$$

$$X(x, y) = (y^2, 0), \quad Y(x, y) = (0, x^2).$$

Each of X, Y has a complete flow, but $X + Y$ does not.

Using the sort of argument in the previous theorem however, one can get a result.

Theorem. Let H be a Hilbert space and let X and Y be locally Lipschitz vector fields which satisfy the following:

  (a) X and Y are bounded and Lipschitz on bounded sets,
  (b) there is a constant $\beta \geq 0$ such that $\langle Y(x), x \rangle \leq \beta \|x\|^2$ for all $x \in H$,
  (c) there is a locally Lipschitz monotone increasing function $c(t) > 0$, $t \geq 0$ such that $\int^\infty \frac{dx}{c(x)} = +\infty$ and $\langle X(x_0), x_0 \rangle \leq \|x_0\| c(\|x_0\|)$

or, stronger, if $x(t)$ is an integral curve of X,

$$\frac{d}{dt}\|x(t)\| \leq c(\|x(t)\|) .$$

Then X, Y and $X + Y$ are positively complete (i.e. complete for $t \geq 0$.)

Note. One may assume $\|X(x_0)\| \leq c(\|x_0\|)$ in (c) instead of $\frac{d}{dt}\|x(t)\| \leq c(\|x(t)\|)$.

Proof. We begin with an elementary comparison lemma:

Lemma. Suppose $r'(t) = c(r(t))$ and $r_0 \geq 0$. Then $r(t) \geq 0$ is defined for all $t \geq 0$. Suppose $f(t) \geq 0$, is continuous and

$$f(t) \leq r_0 + \int_0^t c(f(s))ds , \quad t \in [0, T[.$$

Then

$$f(t) \leq r(t) \quad \text{for} \quad t \in [0, T[.$$

This lemma is not hard to prove. See Hartman [1] for such results.

Proof of Theorem. Let $u(t)$ be an integral curve of $X + Y$. By assumption (a), it suffices to show $u(t)$ is bounded on finite t-intervals, say $t \in [0, T[$. Now using (b),

$$\frac{1}{2}\frac{d}{dt}\|u(t)\|^2 = \langle u(t), X(u(t)) + Y(u(t))\rangle$$

$$\leq \beta\|u(t)\|^2 + \langle u(t), X(u(t))\rangle .$$

By assumption (c) we have for an integral curve $x(t)$ of X,
$\langle x(t), X(x(t)) \rangle = \frac{1}{2} \frac{d}{dt} \|x(t)\|^2 \leq \|x(t)\| c(\|x(t)\|)$. Therefore
$\langle x_0, X(x_0) \rangle \leq \|x_0\| c(\|x_0\|)$ for any $x_0 \in H$. Thus we get

$$\frac{d}{dt} \|u(t)\| \leq \beta \|u(t)\| + c(\|u(t)\|)$$

and hence

$$\frac{d}{dt} (e^{-\beta t} \|u(t)\|) \leq c(\|u(t)\|) .$$

By the lemma, $e^{-\beta t}\|u(t)\|$ is bounded, so $\|u(t)\|$ is bounded. □

9. General Relativity as a Hamiltonian System*

In this lecture we discuss the Einstein field equations of general relativity from the point of view of Hamiltonian systems. In order to motivate the discussion, we digress to include some background and motivational material.

Background.

The basic tenet of special relativity, that the speed of light is constant independent of the movement of source or observer, is reflected in a simple mathematical structure on $R^4 = R^3 \times R$ (space $\times$ time), viz the Minkowski metric:

$$v \cdot w = v^1 w^1 + v^2 w^2 + v^3 w^3 - v^4 w^4$$

or, as a matrix:

$$\begin{pmatrix} +1 & & & 0 \\ & +1 & & \\ & & +1 & \\ 0 & & & -1 \end{pmatrix}$$

(use units such that $c = 1$).

The physically meaningful concepts in special relativity are those invariant under the Lorentz group; i.e. the group of linear isometries of the Minkowski metric.

As Einstein showed in 1905, the above picture - forced by concrete experiments (namely the Michaelson-Morley experiment) - has consequences of a non intuitive nature such as length contractions, time dilatation etc. All this is described in most elementary texts, such as Taylor-Wheeler [1].

---
*This and the next lecture are based on Fisher-Marsden [1,2,5,6].

Later Einstein had the following brilliant insight: it is physically impossible to distinguish gravitational forces from acceleration forces. Indeed, by Galileo's fameous experiment we know that gravitational mass is the same as inertial mass ("principle of equivalence"). But acceleration is a purely geometrical (or kinematical) phenomena. Therefore it should be possible to geometrize space time in such a way that the gravitational fields are part of the geometry itself.

This is what Einstein did in his papers of 1915-17. (See Lanczos [2] for more historical facts).

It is fairly obvious how to generalize Minkowski space. We just use a Lorentz manifold V; i.e. a 4-manifold with a symmetric bilinear form $<,>_\alpha$ on each tangent space $T_x V$, which has signature (+++−).

We want the following to hold: particles in free fall (in the gravitational field) should follow geodesics on V.

Thus we are asserting that a body moving under the force of gravitation alone (e.g. a satellite circling the earth) should travel along a geodesic in an appropriate differentiable manifold. Such a manifold is certainly not flat 3-space, since the motion of a satellite would not then be geodesic. It is also easy to see that the manifold cannot be a curved three-dimensional Riemannian space: consider the case of two projectiles P_1, P_2 launched at the same time from A with trajectories as indicated in the Figure, both passing through B (this is easily arranged). It is clear that not both P_1 and P_2 can be geodesic with respect to any 3-space metric; since B can be moved arbitrarily close to A , there are no normal neighborhoods of A (in which there are unique minimizing geodesics).

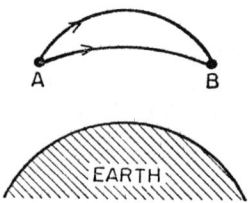

On the other hand, we do get unique trajectories if we require the projectile to pass through B at a given time. So we are compelled to consider a manifold of dimension at least four. Finally, it is almost obvious that this 4-manifold cannot be Riemannian (metric tensor positive definite): Riemannian manifolds are isotropic in the sense that there are no intrinsically defined, distinguished directions. But space time is not isotropic; for example the geodesic connecting (you,now) with (Sirius, 1 second later) could not be traversed by a material particle required to travel at a speed below that of light. One has to distinguish between possible particle trajectories (timelike curves), impossible particle trajectories (spacelike curves), and possible photon trajectories (null curves).

All in all, one is led to consider a four-dimensional Lorentz manifold whose metric tensor g has signature (+++-) . This is quite natural since it tells us that locally (in the tangent space, or in a normal neighborhood) the universe looks like Minkowski-space. As stated above, in this manifold, the "world line" or space time trajectory of a freely falling particle is a geodesic. Furthermore, it is assumed that this geodesic does not depend on the mass of the particle (an orange and a grapefruit behave the same way in the same gravitational field). This is another way of stating the principle of equivalence.

Less obvious than the above is the following. In Newtonian gravitational theory, the gravitational potential φ must satisfy $\nabla^2 \varphi = 0$ exterior to matter. Since the metric is supposed to geometrize these potentials, what conditions should we impose on the metric?

Using this analogy and a good deal of intuition and guesswork, Einstein was led to the (empty space) field equations*

$$R_{\alpha\beta} = 0 .$$

----
* The curvature tensor $R_{\alpha\beta\gamma}{}^{\delta}$ is defined on vector fields by

$$R(X,Y,Z) = \nabla_X \nabla_Y Z - \nabla_Y \nabla_X Z - \nabla_{[X,Y]} Z \quad \text{where} \quad (\nabla_X Y)^{\alpha} = X^{\beta} \frac{\partial Y^{\alpha}}{\partial x^{\beta}} + \Gamma^{\alpha}_{\beta\gamma} X^{\beta} Y^{\gamma}$$

(summation on repeated indices) and

$$\Gamma^{\alpha}_{\beta\gamma} = \tfrac{1}{2} g^{\alpha\rho} \left(\frac{\partial g_{\beta\gamma}}{\partial x^{\rho}} - \frac{\partial g_{\rho\beta}}{\partial x^{\gamma}} - \frac{\partial g_{\rho\gamma}}{\partial x^{\beta}} \right)$$

Here $g_{\alpha\beta}$ is the components of the metric in a chart, and $g^{\alpha\rho}$ is the inverse matrix. We write $x^{\alpha} = (x^j, t)$ for local coordinates. Also raising indices corresponds to identifying $T_x V$ and $T_x^* V$ via g, as usual (see lecture 2) e.g.: $X_{\beta} = g_{\beta\gamma} X^{\gamma}$ etc.

The Ricci tensor Ric = $R_{\alpha\beta}$ is $R_{\alpha\beta} = R_{\alpha\delta\beta}{}^{\delta}$ (a contraction), and the scalar curvature is $R = R_{\alpha}{}^{\alpha}$. One writes $A_{\alpha\beta|\gamma}$ for the covariant derivative of a tensor.

Following this, Einstein incorporated matter, or other external sources via

$$G_{\alpha\beta} \equiv R_{\alpha\beta} - \tfrac{1}{2} R g_{\alpha\beta} = T_{\alpha\beta}$$

where $T_{\alpha\beta}$ is a given energy momentum tensor of the sources ($T_{\alpha\beta}$ is divergence free and $R_{\alpha\beta}$ is not, so Einstein modified $R_{\alpha\beta}$ to $G_{\alpha\beta}$ - fortunately $R_{\alpha\beta} = 0$ is equivalent to $G_{\alpha\beta} = 0$).

Sometimes a "cosmological constant" λ is also included:

$$G_{\alpha\beta} - \lambda g_{\alpha\beta} = T_{\alpha\beta} .$$

Later these equations were "justified" on theoretical grounds by Cartan-Weyl. They proved that any symmetric divergence free 2-tensor depending on g, Dg, $D^2 g$ had to have the form* $\mu G_{\alpha\beta} - \lambda g_{\alpha\beta}$.

There is another piece of motivation that lends insight into the nature of the field equations which is due to Pirani [1]. This proceeds as follows:

Let u be the tangent to a timelike geodesic $x(t)$ (timelike means $<u,u> < 0$), so $\nabla_u u = 0$. Consider the Jacobi field (or deviation vector) η along $x(t)$; it satisfies Jacobi's equation:

$$\nabla_u \nabla_u \eta + R(\eta, u) u = 0$$

where R is the curvature tensor. Regarded as a map $\tilde{R}_u$ in η, $Ric(u,u)$ is its trace.

We are supposing $Ric = 0$. Let e_i, $i=1,2,3$ be vectors

---
* It is usually assumed that the tensor depends linearly on $D^2 g$, but see Rund-Lovelock [1].

orthogonal to u at a point p where $t = 0$. Then extend e_i to be Jacobi fields with initial condition $\nabla_u e_i = 0$ at p. Then

$$\nabla_u(u \cdot e) = \nabla_u u \cdot e_i + u \cdot \nabla_u e_i = u \cdot \nabla_u e_i \text{ so } \nabla_u \nabla_u(u \cdot e_i) = u \cdot \nabla_u \nabla_u e_i = -u \cdot R(e_i, u)u = 0$$

(we have $< R(v,u)U, u > = 0$ always by skew symmetry of $< R(u,v)w, z >$ in w, z). Hence $u \cdot e_i = 0$ for all time. Choose e_i to be eigenvectors of $\tilde{R}_u$ on the space orthogonal to u. We denote this restriction by $\tilde{R}_u^\perp$. Thus $\tilde{R}_u^\perp e_i = \lambda_i e_i$, and $\lambda_1 + \lambda_2 + \lambda_3 = 0$ because $\tilde{R}_u$ is zero.

Now these vectors e_i span a three volume. Multiply e_i by ε (so that we can be sure exp maps the field εe_i onto geodesics close to the geodesic through p). A satisfactory approximation of the volume of the cube spanned by these vectors is $(\text{vol}) = \varepsilon^3 e_1 \wedge e_2 \wedge e_3$. We compute:

$$^4g_{\alpha\beta} dx^\alpha dx^\beta = -dt^2 + g_{ij} dx^i dx^j \ .$$

$$\frac{d^2}{ds^2}(\text{vol})\Big|_p = \nabla_u \nabla_u (\text{vol})\Big|_p$$

$$= \varepsilon^3 \{\nabla_u \nabla_u e_1) \wedge e_2 \wedge e_3 + e_1 \wedge (\nabla_u \nabla_u e_2) \wedge e_3 + e_1 \wedge e_2 \wedge \nabla_u \nabla_u e_3$$

$$+ \text{First derivative terms}\}\Big|_p \ .$$

Since the e_i's are Jocobi fields and eigenvectors (at p) of $\tilde{R}_u^\perp$, and since $\nabla_u e_i\big|_p = 0$, we have

(1) $\quad \dfrac{d^2}{dt^2}(\text{vol})\Big|_p = -(\lambda_1 + \lambda_2 + \lambda_3)(\text{vol})\Big|_p = 0$ if $\dfrac{d(\text{vol})}{dt}\bigg|_p = 0$

as the condition equivalent to Ric = 0 .

We can interpret this more physically as follows. Imagine ourselves in a freely falling elevator and watch a collection of freely falling particles. The particles are initially at rest with respect to each other, but due to motion towards the earth's center, they will pick up a relative motion (see the following Figure).

The condition (1) says that the 3-volume (up to second order) is remaining constant during the motion. This geometric property is directly verifiable in the case of the Newtonian gravitational field, so is a reasonable condidate for generalization. Thus we shall adopt Ric = 0 as the Einstein field equations in our Lorentz four manifold.[*]

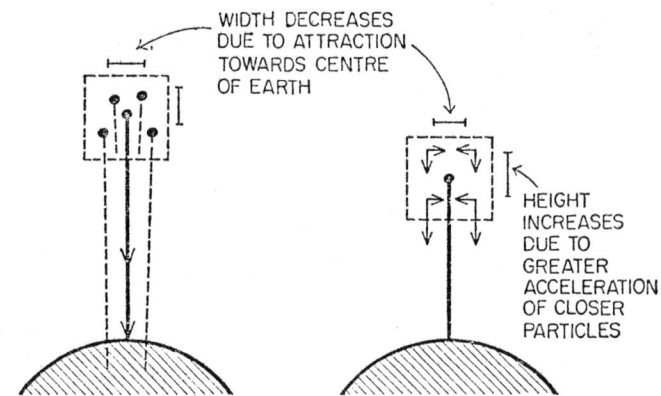

The general program

Let V be a spacetime with M a three-dimensional spacelike section without boundary (a spacelike section is a submanifold such that for $0 \neq v \in T_x M, <v,v> \; > 0$) . Assume for the moment that M is compact, so that there exists a neighborhood U of M in which the timelike geodesics (that is geodesics whose tangent vectors v have

---

[*] See also J. Wheeler [2].

$<v,v> < 0$) orthogonal to M have no focal points. If we let t measure proper time on these geodesics, with $t = 0$ on $M = M_0$, then the function t is well-defined in U. The surfaces M_t given by $t =$ constant form a one parameter family of space sections, all diffeomorphic to M. Let g_t be the induced Riemannian metric on M_t. Via the aformentioned diffeomorphism, we can regard g_t as a curve in the space of positive definite metrics on M. The fact that V is Ricci flat implies that g_t satisfies certain differential equations. We want to work these out.

We also want to go the other way: given M, a positive definite metric g_0, and a symmetric tensor $k_0 = \dot{g}_0$ (the second fundamental form of M in V) we want to find the curve g_t describing the time evolution of the geometry of M, and then to past together the resulting 3-manifolds M_t to obtain a piece of spacetime.

The Space of Riemannian Metrics.

Fix a 3-manifold M which we shall take to be compact for simplicity. This is supposed to represent a model for the spatial universe. Let $S_2(M)$ be the set of all C^∞ symmetric two tensors on M and let $\mathfrak{m} \subset S_2(M)$ be the cone of positive definite ones; i.e. riemannian metrics.

The "time evolution" of the universe will be represented by

a curve of metrics $g(t) \in \mathcal{M}$. Of course there is no global time scale and physically this evolution takes place relative to a given system of "clocks" and a global "frame". This point will be briefly discussed below. We want $\mathcal{M}$ to be the configuration space for a dynamical system. The first job therefore will be to construct a metric on $\mathcal{M}$.

As was the case with hydrodynamics, one should properly work with metrics in the Sobolev class H^s. This space is denoted $\mathcal{M}^s$. For simplicity most of the development will be done in $\mathcal{M}$. Below we shall discuss briefly the existence questions, which make use of H^s. This also comes into play in lecture 10. Since $\mathcal{M}$ is an open cone in $S_2(M)$ (using the H^s or C^∞ topology), for $g \in \mathcal{M}$, we have

$$T_g\mathcal{M} = S_2(M)$$

so
$$T\mathcal{M} = \mathcal{M} \times S_2(M)$$

Define a metric $\mathcal{G}$ on $\mathcal{M}$ as follows:

$$\mathcal{G}_g(h,k) = \int_M \{h \cdot k - (\mathrm{tr}\,h)(\mathrm{tr}\,k)\} d\mu_g \tag{1}$$

where $g \in \mathcal{M}$, $h,k \in T_g\mathcal{M}$, $h \cdot k$ is the induced inner product, $h \cdot k = h^{ij}k_{ij}$, $\mathrm{tr}\,h = h^i_i$ is the trace and μ_g is the volume determined by g. Observe that $h \cdot k$ and $\mathrm{tr}\,h$ both depend upon g. Thus $\mathcal{G}$ is a non constant metric. $\mathcal{G}$ is called the <u>deWitt metric</u>. Although $\mathcal{G}$ is not positive definite, we can easily demonstrate that $\mathcal{G}$ is weakly non degenerate: suppose $\mathcal{G}_g(h,k) = 0$ for all $h \in S_2$.

Then $\mathcal{G}_g(k, k - \frac{1}{2}(trk)g) = 0$. But this equals $\int_M k \cdot k$ so $k = 0$.

The first thing we will want to do is work out the spray of $\mathcal{G}$.

Proposition. <u>The spray</u> S <u>of</u> $\mathcal{G}$ <u>is given as follows:</u> <u>Its principal part is</u>

$$S: T\mathbb{M} \to S_2 \times S_2$$

$$S(g,k) = (k, k \times k - \frac{1}{2}(trk)k - \frac{1}{8}\{k \cdot k - (trk)^2\}g) \qquad (2)$$

Note. The factor $\frac{1}{8}$ here depends on $\dim M = 3$.

Proof. We use the formulas for the spray in lecture 2. We first must compute the derivative of $\mathcal{G}_g$ with respect to g. We do this in three steps:

<u>Lemma.</u> <u>The derivative of</u> $g \mapsto \mu_g$ <u>in direction</u> $h \in S_2$ <u>is given by</u> $\frac{1}{2}(trh)\mu_g$.

Proof. Let $g(t) = g + th$. The derivative in question is $\frac{d}{dt} \mu_{g(t)}\big|_{t=0}$. Using the local formula $\mu_g = \sqrt{\det g_{ij}}\, dx^1 \wedge \ldots \wedge dx^n$ we get the result from the formula $\frac{d}{dt} \det(g_{ij} + th_{ij})\big|_{t=0} = tr(h_{ij})\det(g_{ij})$.

The latter formula may be proven from the fact that the derivative det at the identity is trace, so $\frac{d}{dt} \det(g_{ij} + t \cdot h_{ij}) = \det(g_{ij}) \frac{d}{dt} \det(1 + tg^{-1}h)$

$= \det(g_{ij})\, tr(h_{ij})$. Note $g: T_xM \to T_x^*M$ so $g^{-1}h$ is a linear map from T_xM to itself and trh is the trace of this map. In coordinate language $g^{-1}h$ raises one index on h. □

Lemma. *The derivative of* $g \mapsto h \cdot k$ *in direction* h_1 *is given by* $-2 h_1 \cdot (h \times k)$ *where* $h \times k = hg^{-1}k$ *or* $(h \times k)_{ij} = h_{i\ell} k^{\ell}{}_j$, *in coordinates.*

Proof. Now $h \cdot k = \text{tr}(g^{-1} h g^{-1} k)$ and as usual, $\frac{d}{dt} g(t)^{-1} = -g^{-1} h_1 g^{-1}$, where $g(t) = g + th_1$. Thus we get for the derivative

$$- \text{tr}(g^{-1} h_1 g^{-1} h g^{-1} k) - \text{tr}(g^{-1} h g^{-1} h_1 g^{-1} k)$$

and this gives the result. □

In a similar way, one proves

Lemma. *The derivative of* $g \mapsto \text{tr}(h)$ *in direction* h_1 *is given by* $-h_1 \cdot h$.

Continuing with the proof the proposition, we have from lecture 2 that if we write $S(g,k) = (k, S_g(k))$, S_g should satisfy:

$$G_g(S_g(k),h) = \tfrac{1}{2} D_g \{G_g(k,k)\} \cdot h - D_g G_g(k,h) \cdot k \qquad (3)$$

From the lemmas we get

$$D_g G_g(k,h) \cdot h_1 = \int_M (-2h_1 \cdot (k \times h) + (\text{tr}h) h_1 \cdot k + (\text{tr}k) h_1 \cdot h)\, d\mu_g$$

$$+ \int_M [h \cdot k - (\text{tr}h)(\text{tr}k)] \cdot \tfrac{1}{2} \text{tr} h_1 \, d\mu_g .$$

Thus the right hand side of (3) becomes

$$\int_M \{h \cdot (k \times k) - \tfrac{1}{2} h \cdot k (trk) - (trh) k \cdot k + \tfrac{1}{2}(trk)^2 trh + \tfrac{1}{4}[k \cdot k - (trk)^2] trh\} d\mu_g \quad (4)$$

while the left side is $\int_M \{S_g(k) \cdot h - tr(S_g k) tr(h)\} d\mu_g$ which becomes,

on substituting the stated expression for S_g, using $g \cdot h = trh$,

$trg = 3$, and $tr(k \times k) = k \cdot k$,

$$\int_M h \cdot (k \times k) - \tfrac{1}{2}(h \cdot k) trk - \tfrac{1}{8} \{k \cdot k \, trh - (trk)^2 trh\} d\mu_g$$

$$- \int_M k \cdot k \, trh - \tfrac{1}{2}(trk)^2 trh - \tfrac{1}{8} \{k \cdot k \cdot 3 \, trh - (trk)^2 3 trh\} d\mu_g$$

which equals (4) above. $\square$

The Gravitational Potential.

We have established our metric $\mathcal{G}$ on $\mathfrak{M}$ and have determined its spray. We now proceed to consider a potential and will compute its gradient. The spray S of $\mathcal{G}$ is simply algebraic, whereas the gradient of the potential will involve non linear differential operators.

Define $V: \mathfrak{M} \to R$ by $V(g) = 2 \int_M R(g) d\mu_g$ where $R(g)$

is the scalar curvature (remember g is a three dimensional metric) and as usual μ_g is the volume associated with g.

Proposition. *The gradient of* V *with respect to the metric* $\mathcal{G}$ *on* $\mathfrak{M}$ *is*

$$grad \, V(g) = -2 \, Ric(g) + \tfrac{1}{2} R(g) g \in S_2(M) \quad .$$

Proof. Let $g(t) = g + th$. Then

$$dV(g) \cdot h = 2 \frac{d}{dt} \int_M R(g(t)) \, d\mu_{g(t)} \Big|_{t=0}$$

The derivative may be done in two parts. The $\mu_{g(t)}$ part is taken care of by the lemmas. For the scalar curvature we use:

Lemma. $\frac{d}{dt} R(g(t)) \Big|_{t=0} = \Delta(\mathrm{tr}\,h) + \delta\delta h - \mathrm{Ric}(g) \cdot h$

where $\delta\delta h = h^{ij}{}_{|i|j}$ is the double covariant divergence.

This is a straightforward but somewhat lengthy computation which we shall omit. (See Lichnerowicz [2]).

Since we are taking M compact with no boundary, the two terms $\Delta(\mathrm{tr}\,h)$, $\delta\delta h$ drop out by Stokes theorem. Hence we get

$$dV(g) \cdot h = -2 \int_M \mathrm{Ric}(g) \cdot h \, d\mu_g + \int_M R(g) \mathrm{tr}(h) \, d\mu_g \,,$$

It is now easy to verify that the formula in the proposition satisfies

$$\mathcal{G}_g(\mathrm{grad}\, V(g), h) = dV(g) \cdot h$$

if we remember that $\mathrm{tr}(\mathrm{Ric}(g)) = R(g)$ and $\mathrm{tr}\,g = 3$. $\square$

The Energy Condition; Coordinate Invariance and Conservation Laws

If we consider the Lagrangian $L(g,k) = \frac{1}{2} \mathcal{G}_g(k,k) - V(g)$ on $\mathfrak{m}$, then we have computed above the corresponding spray to be $S_g(k) - \mathrm{grad}\, V(g)$. Thus an integral curve $g(t)$ satisfies

$$\frac{d^2 g}{dt^2} = S_g\left(\frac{dg}{dt}\right) - \mathrm{grad}\, V(g) \,.$$

These equations have an important property which is not shared by the usual non relativistic field theories. This is that not only is the total energy conserved, but it is pointwise conserved. Actually this law is intimately connected with another conservation law which we shall develop first.

Theorem. <u>Let $\pi = ((\mathrm{tr}\,k)g - k) \otimes \mu_g$, the conjugate momentum. Then along an integral curve of L above, $\frac{\partial}{\partial t}\delta\pi = 0$. Here, $\delta\pi$ is defined by $\delta((\mathrm{tr}\,k)g - k) \otimes \mu_g$; $\delta h = h_{ij}{}^{li}$. In particular if</u> $\delta\pi = 0$ at $t = 0$, <u>then this condition is maintained. Furthermore, this law is the conservation law associated with the invariance of</u> L <u>under the (left) action of the group of diffeomorphisms</u> $\mathfrak{D}$ <u>on</u> $\mathfrak{M}$ <u>by</u> $\Phi_\eta(g) = \eta_* g$.

In otherwords, we get a free conservation law just because our theory is invariant under coordinate transformations. The actual form of V is irrelevant.

Proof. We are considering the action of $\mathfrak{D}$ on $\mathfrak{M}$ as stated. See lecture 4 for the relevant properties of $\mathfrak{D}$ which are used here. Consider X a vector field on M, so X is in the Lie algebra of $\mathfrak{D}$. The one parameter subgroup corresponding to X is its flow $F_t \in \mathfrak{D}$. Since

$$\frac{d}{dt} F_{t*}g \Big|_{t=0} = -L_X g \;,$$

we see that the corresponding infinitesimal generator on $\mathfrak{M}$ is

$g \mapsto - L_X g \in S_2(M)$. Hence by our conservation laws (lecture 6),

$$(g,k) \mapsto \mathcal{G}_g(k, -L_X g)$$

is a conserved quantity. At this point, we need the following:

Lemma. $\int_M L_X g \cdot k \, d\mu_g = -2 \int_M X \cdot \delta k \, d\mu_g$.

Proof. It is easy to derive the following formula $L_X g = X^i{}_{|j} + X^j{}_{|i}$.

From this it follows that $\delta(k \cdot X) = (\delta k) \cdot X + k \cdot \nabla X = (\delta k) \cdot X + \frac{1}{2} k \cdot L_X g$.

Since, by Stokes theorem $\int_M \delta(k \cdot X) \, d\mu = 0$, we get the lemma. □

Now
$$\mathcal{G}_g(k, L_X g) = \int_M \{k \cdot L_X g - (\operatorname{tr} k)(\operatorname{tr} L_X g)\} d\mu$$

$$= \int_M (L_X g) \cdot (k - (\operatorname{tr} k) g) \, d\mu$$

$$= \int_M L_X g \cdot \pi$$

$$= -2 \int_M X \cdot \delta \pi$$

Thus for any vector field X, $\int_M X \cdot \delta \pi$ is conserved. Hence $\delta \pi$ itself is conserved. □

This result could also be obtained from Noethers theorem. Notice that the bundle in question is $S^2(M)$, and since $\mathcal{L}$ depends on second derivatives of the fields g, since $R(g)$ does, one would have to use the second jet bundle. That approach seems more complicated.

The energy conservation law is as follows.

Theorem. For the equations for L above, we have

$$\frac{\partial}{\partial t} \{\mathcal{H}(g,k)\, \mu_g\} + 2\, \delta\delta\pi = 0$$

where

$$\mathcal{H}(g,k) = \tfrac{1}{2}\, \hat{G}_g(k,k) + 2\, R(g)$$

is the energy density. In particular if $\delta\pi = 0$, $\mathcal{H} = 0$ at $t = 0$ then these conditions are maintained in time.

Proof. Let $\mathcal{K}_g(k) = \tfrac{1}{2}\{k \cdot k - (\mathrm{tr}\,k)^2\}$ the kinetic energy density.

Then

$$\frac{\partial}{\partial t}(\mathcal{K}\mu_g) = \hat{G}_g(k, \frac{dk}{dt})\, \mu_g + D_g \mathcal{K}_g(k) \cdot k\, \mu_g$$

$$+ \frac{\mathcal{K}}{2}\, \mathrm{tr}(k)\, \mu_g$$

where $\hat{G}$ is the pointwise Dewitt metric. Using the lemmas on p. 213-214,

$$D_g \mathcal{K}_g(k) \cdot k = - k \cdot (k \times k) + k \cdot k (\mathrm{tr}\,k)$$

and

$$\hat{G}_g(k, \frac{dk}{dt}) = k \cdot k \times k = \tfrac{1}{2}(\mathrm{tr}\,k) k \cdot k - \frac{1}{4}\mathcal{K}(\mathrm{tr}\,k)$$

$$- \{\mathrm{trk}(k \cdot k) - \tfrac{1}{2}(\mathrm{tr}\,k)^3 - \frac{3}{4}\mathcal{K}(\mathrm{tr}\,k)\}$$

$$+ 2\, \mathrm{Ric}(g) \cdot k - 2\, R(g)\, \mathrm{tr}(k)$$

$$- \tfrac{1}{2}\, r(g)\, \mathrm{tr}(k) + \frac{3}{2}\, R(g)\, \mathrm{tr}(k)$$

Adding we get

$$\frac{\partial}{\partial t}(\mathcal{K}\mu_g) = 2\, \mathrm{Ric}(g) \cdot k - R(g)\, \mathrm{tr}(k)\ .$$

On the other hand,

$$\frac{\partial}{\partial t}(2 R(g) \mu_g) = R(g)\mathrm{tr}(k)\mu_g + (2\Delta(\mathrm{tr}k)$$

$$+ 2\delta\delta k - 2\,\mathrm{Ric}(g)k)\mu_g\ .$$

Hence adding,

$$\frac{\partial}{\partial t}(\mathcal{H}\mu_g) = 2(\Delta(\mathrm{tr}k) - \delta\delta k)\mu_g$$

$$= -2\delta\delta\pi\ .\ \ \square$$

There is good a priori evidence, including a theorem, that any genuinely relativistic theory, in the absense of external fields, must have $\mathcal{H}$ pointwise constant (see Fischer-Marsden [1]).

Therefore one selects out the subset C of $T\mathfrak{M}$ defined by:

$$C = \{(g,k)\mid \tfrac{1}{2}\{k\cdot k - (\mathrm{tr}k)^2\} + 2R(g) = 0 \text{ and}$$

$$\delta(k - (\mathrm{tr}k)g) = 0\} \tag{5}$$

The previous results prove that our Hamiltonian flow on $T\mathfrak{M}$ leaves C invariant and we thus select out C as the physically meaningful subset. It is rather analogous to what one does in electromagnetism. In general, C is not a manifold. This point is discussed in lecture 10.

Thus for $(g,k) \in C$, the evolution equations become,

$$\left.\begin{aligned}\frac{\partial g}{\partial t} &= k \\[4pt] \frac{\partial k}{\partial t} &= S_g(k) - \mathrm{grad}\ V(g) \\ &= k \times k - \tfrac{1}{2}(\mathrm{tr}k)k + 2\,\mathrm{Ric}(g)\end{aligned}\right\} \tag{6}$$

The extra terms have dropped out in view of $H = 0$.

Relationahip with the Four Geometry.

We now form $L = M \times R$ and construct a Lorentz metric on L as follows.

$$\tilde{g}_{(x,t)}((v,r),(w,s)) = g_x(t)(v,w) - rs$$

where $(v,r), (w,s) \in T_{(x,t)}(M \times R) \approx T_xM \times R$ and g is the time dependent metric on M. In coordinates, the formula reads:

$$\tilde{g}_{\alpha\beta} dx^\alpha dx^\beta = g_{ij} dx^i dx^j - dt^2$$

where $x^\beta = (x^i, t)$; $i = 1,2,3$, $\beta = 0,1,2,3$.

Theorem. <u>The Lorentz metric $\tilde{g}$ is Ricci flat if and only if g satisfies the evolution equations (6) above, together with the initial constraints (5).</u>

This result therefore establishes the equivalence between solving the intiial value problem for the three metric g and Ricci-flatness of the four metric $\tilde{g}$ i.e. the Einstein field equations. Note that we have taken a special form for $\tilde{g}$, namely we have assumed $\tilde{g}_{0i} = 0$, $\tilde{g}_{00} = -1$. This point is discussed below.

The proof turns on the Gauss-Codazzi equations which relate the curvatures on L,M with the second fundamental form and the unit normal. This result which we assume here, is the following, for the case at hand:

Lemma. Let S_{ij} be the second fundamental form on M and Z^μ the unit normal to M, so $Z^\mu = (0,1)$. Let $^{(4)}R_{\alpha\beta\gamma\delta}$ be the curvature tensor on L, $^{(3)}R_{ijk\ell}$ that on M. Then

(i) $\quad ^{(4)}R_{0i0j} = \dfrac{\partial S_{ij}}{\partial t} - (S \times S)_{ij}$

(ii) $\quad ^{(4)}R_{ijk\ell} = {}^{(3)}R_{ijk\ell} + S_{ik}S_{j\ell} - S_{i\ell}S_{jk}$

(iii) $\quad ^{(4)}R_{0ijk} = S_{ik|j} - S_{ij|k}$

Now if $g, \tilde{g}$ are related as before, we assert that

$$S_{ij} = -\tfrac{1}{2} k_{ij}$$

where $k_{ij} = \dfrac{\partial g}{\partial t}$. Indeed, we have $S_{ij} = -Z_{i|j} = -\Gamma^\alpha_{ij} Z_\alpha = -\Gamma^0_{ij}$.
But from the formula for the Christoffel symbols, we compute that $\Gamma^0_{ij} = \tfrac{1}{2} g_{ij,0}$, and so our claim holds.

Now suppose $\tilde{g}$ is Ricci flat. Then in particular,

$$0 = {}^{(4)}R_{ij} = \tilde{g}^{\alpha\beta} \, {}^{(4)}R_{\alpha i j \beta} = \tilde{g}^{00} \, {}^{(4)}R_{0ij0} + \tilde{g}^{k\ell} \, {}^{(4)}R_{kij\ell} .$$

Applying (i), (ii) of the lemma with $S_{ij} = -\tfrac{1}{2} k_{ij}$ gives

$$0 = -\tfrac{1}{2} \dfrac{\partial k_{ij}}{\partial t} + \tfrac{1}{4}(k \times k)_{ij} + g^{k\ell}\{{}^{(3)}R_{kij\ell} + \tfrac{1}{4}(k_{kj} k_{i\ell} - k_{k\ell} k_{ij})\}$$

or $\quad \dfrac{\partial k_{ij}}{\partial t} = \tfrac{1}{2}(k \times k)_{ij} + 2{}^{(3)}R_{ij} + \tfrac{1}{2}(k \times k)_{ij} - \tfrac{1}{2}(\text{trk})k$

$$= (k \times k)_{ij} - \tfrac{1}{2} k_{ij}(\text{trk}) + 2{}^{(3)}R_{ij}$$

which is the correct equation of motion for k, according to (6). Similarly from

$$0 = {}^{(4)}R_{oi} = \tilde{g}^{\alpha\beta}\,{}^{(4)}R_{\alpha o i \beta} = -{}^{(4)}R_{ooio} + \tilde{g}^{k\ell}\,{}^{(4)}R_{koi\ell}$$

$$= \tilde{g}^{k\ell}\,{}^{(4)}R_{koi\ell}$$

we obtain from (iii)

$$0 = \tfrac{1}{2} g^{k\ell} \{k_{k\ell|i} - k_{ki|\ell}\}$$

or $\quad\quad \delta(\text{trk } g - k) = 0$

Similarly from ${}^4R_{oo} = 0$ we obtain the energy statement. The converse is proved in exactly the same way. $\square$

The Lapse and Shift.

Although any Lorentz metric $\tilde{g}$ can be put in the form $\tilde{g} = g_{ij}\,dx^i dx^j - dt^2$ by a suitable coordinate change (namely in gaussian, or normal, coordinates), the above description is incomplete since it singles out this coordinate system as special. The situation can be remedied however, by introducing what are called the lapse and shift functions. The shift function is a time dependent vector field X prescribed in advance, corresponding to a choice of coordinate system. Now we set

$$\tilde{g}_{\alpha\beta} = -(1 - <X,X>)dt^2 - 2X_i dt\, dx^i + g_{ij} dx^i dx^j$$

and this corresponds to the evolution equations

$$\frac{\partial g}{\partial t} = k - L_X g$$

$$\frac{\partial k}{\partial t} = S_g(k) - 2\,\text{Ric}(g) - L_X k$$

This can all be seen very simply as a change from what we might call "space" to "body" coordinates. Namely, if $\bar{g}$ is is a solution for no shift and η_t is the flow of X, then $g = (\eta_t^{-1})^* \bar{g}$ solves the above. $\tilde{g}$ above is just the metric in the induced coordinate change on $M \times R$.

This therefore takes care of coordinate changes on $M \times R$ corresponding to changes in M. For changes along R one introduces the lapse and things now become more involved. We now introduce $N: M \times R \to R$ and

$$\tilde{g}_{\alpha\beta} = -N dt^2 + g_{ij} dx^i dx^j$$

with

$$\begin{cases} \dfrac{\partial g}{\partial t} = Nk \\[2mm] \dfrac{\partial k}{\partial t} = NS_g(k) - 2N\,\text{Ric}(g) + 2\,\text{Hess}\,N \end{cases}$$

where $\text{Hess}\,N = N_{|i|j}$ is the Hessian of N. We shall not go into details here except to remark that this can be handled by the following device. Set $\mathcal{T}$ = the C^∞ maps ξ ; $M \times R \to R$. If one knows how to treat relativistic particles by extension of the Lagrangian to a homegeneous degenerate one, (see Lancos [1]) then we are motivated to extend our Lagrangian from $\mathbb{m}$ to $\mathbb{m} \times \mathcal{T}$ with $\mathcal{T}$ generalizing a single time parameter R . This procedure leads to the above equations of motion, with the degeneracy reflected in the arbitrariness of N . One can use the symmetry groups $\mathcal{D}$ and $\mathcal{T}$ to construct a reduced phase space using the methods of lecture 6 to recover a result of Fadeev [1]. See Marsden-Fischer [1]. However we shall not pursue the matter further here.

Remarks on existence of solutions

The original theorem concerning existence of solutions for the Einstein system is due to Fourès-Bruhat [1]. The result was improved on by Lichnerowicz [1] using Leray systems. See also Choquet-Bruhat [1] and Dionne [1]. The method involves the theory of second order partial differential equations which are quasi-linear and "strictly hyperbolic". Actually, there is a simpler theory of quasi-linear first order systems which is applicable here (cf. Fischer-Marsden [2,3]).

The way this goes is a bit complicated, and will not be presented in detail here. We will illustrate with the wave equation how one reduces a second order system to a first order one. The method for relativity is more complicated, but the basic idea is the same.

First of all let us consider the linear problem in R^n: Let u be a vector-valued function $u:R^n \to R^m$. The system

(1) $$\frac{\partial u}{\partial t} = \sum_{i=1}^{n} A^i(x) \frac{\partial u}{\partial x^i} + B(x) \cdot u$$

is said to be <u>symmetric hyperbolic</u> if the $m \times m$ matrices A^i are symmetric for all $1 \leq i \leq n$. The system is first order and linear in u. Under fairly mild restrictions (A^i, B should be of class H^s, $s > (n/2) + 1$), there exists a unique solution u_t in H^s (all time) for any initial condition u_0 in H^s. This result is due basically to Petrovsky [1], Friedrichs [1], and others. A proof may be found in Courant-Hilbert [1] Vol. II; see also Kato [3,4] and Dunford-Schwartz [1]. Using standard techniques of reducing second order systems to first order, this theorem may be used to solve the wave equation in R^n:

EXAMPLE. The wave equation.

The equation is

$$\frac{\partial^2 f}{\partial t^2} = \nabla^2 f; f = f(x^1,\ldots,x^n,t) \,.$$

Put, formally,

$$\begin{bmatrix} f \\ \frac{\partial f}{\partial x^1} \\ \vdots \\ \frac{\partial f}{\partial x^n} \\ \frac{\partial f}{\partial t} \end{bmatrix} = \begin{bmatrix} u_0 \\ u_1 \\ \vdots \\ u_n \\ u_{n+1} \end{bmatrix}$$

Then the wave equation for f is the same as the following symmetric hyperbolic system for u :

$$\begin{cases} \dfrac{\partial u_0}{\partial t} = u_{n+1} \\[2mm] \dfrac{\partial u_1}{\partial t} = \dfrac{\partial u_{n+1}}{\partial x^1} \\[2mm] \vdots \\[2mm] \dfrac{\partial u_n}{\partial t} = \dfrac{\partial u_{n+1}}{\partial x^n} \\[2mm] \dfrac{\partial u_{n+1}}{\partial t} = \dfrac{\partial u_1}{\partial x^1} + \ldots + \dfrac{\partial u_n}{\partial x^n} \end{cases}.$$

In this case

$$A^1 = \begin{bmatrix} 0 & 0 & \ldots & 0 \\ 0 & 0 & \ldots & 1 \\ 0 & 0 & \ldots & 0 \\ \vdots & & & \\ 0 & 1 & \ldots & 0 \end{bmatrix} \quad \text{etc.}$$

are symmetric $(n+2) \times (n+2)$ matrices.

Thus, using the linear theory for general first order symmetric hyoerbolic systems, we get an existence theorem for the wave equation, namely that if $(f_0, (\partial f_0/\partial t)) \in H^{s+1} \times H^s$ there is a unique solution $f_t \in H^{s+1}$, $-\infty < t < \infty$, satisfying the given initial conditions.

The hyperbolicity of $(\partial^2 f/\partial t^2) = \Delta^2 f$ is reflected in the symmetry of the A^i. If we had used $(\partial^2 f/\partial t^2) = -\Delta^2 f$, the A^i would not have come out symmetric - the Cauchy problem in this case is not well posed.

Now consider the nonlinear problem in R^n. In this case we have a system of the form

$$\frac{\partial u}{\partial t} = \sum_i A^i(x,t,u) \frac{\partial u}{\partial x^i} + B(x,t,u) ,$$

where the A^i and B are matrices which are polynomial in u (or more generally, satisfy Sobolev's "condition T"; cf. Sobolev [1]). The system is quasi-linear and the matrices A^i are symmetric. The nonlinear theorem is obtained from the linear theory by adapting the Picard method. In this case also, unique solutions exist in H^s, but only for short time, in contrast to the linear theory.

The Einstein system above is rather like the wave equation and one can show that in the appropriate variables, obtained in a way not unlike that for the wave equation, it is symmetric hyperbolic.

The verification that it is symmetric hyperbolic uses "harmonic coordinates"; cf. Lichnerowicz [1].

Thus we get existence and uniqueness of smooth solutions for short time (which can be extended to maximal solutions as well). These solutions depend continuously on the initial data.

For details of all of this, see Fischer-Marsden [2].

10. **Linearization Stability of the Einstein Equations.**[+]

This lecture is concerned with some "hard" applications of global analysis methods to general relativity.

There have been a number of impressive applications of differential topology to relativity recently. One of the most important of such applications has been to the study of the topology of spacetimes in the works of Geroch, Hawking and Penrose. Using techniques of differential topology and differential geometry, they prove, for example, various incompleteness theorems from which one may infer the existence of black holes -- under reasonable mathematical hypotheses on the spacetime involved. See W. Kundt [1] for a recent survey and a bibliography for this subject.[*]

The techniques used in the above are taken from the study of the topology and geometry of <u>finite</u> dimensional manifolds. Our main concern here is with the applications of <u>infinite</u> dimensional manifolds.

That infinite dimensional manifold theory is relevant for general relativity was first pointed out by J. A. Wheeler [1]. He stressed the usefulness of considering superspace S. S consists of riemannian metrics on a given three manifold M, with metrics which can be obtained one from the other by a coordinate transformation, identified, i.e. $S = \mathcal{M}/\mathcal{D}$. This space S is important for we can view the universe as an evolving (or time dependent) geometry and hence as a curve in S as explained in lecture 9. The geometry and

---
[*] See also the new book of Hawking and Ellis, Cambridge (1973).
[+] Part of this lecture is adapted from the 1973 essay of A. Fischer and J. Marsden in the Gravity Research Foundation.

topology of S has been investigated by several people. See for example Fischer [1].

Recall from lecture 9 that the Einstein field equations state that, outside of regions of **matter**, the metric tensor $g_{\alpha\beta}$ must be Ricci flat; i.e. $R_{\alpha\beta} = 0$. (1) This is a complicated coupled system of non-linear partial differential equations. One can regard the Einstein equations as a Hamiltonian system of differential equations on S in an appropriate sense. This idea goes back to Arnowitt, Deser and Misner [1] but was put into the setting of S, explicitly using infinite dimensional manifolds by Fischer-Marsden [1].

The above applications to general relativity can be regarded as "soft" in the sense that infinite dimensional manifolds are involved mostly as a language convenience and as a guide to the theory's structure. While this is important, it is perhaps not critical to the development of the theory.

The first substantial "hard" theorem using infinite dimensional analysis (at least in an informal way) is due to Brill and Deser [1]. They establish the important result that any non-trivial perturbation of Minkowski space leads to a spacetime with strictly positive mass (or internal gravitational energy). The technique they use is an adaptation of methods from the calculus of variations. The idea behind the proof is rather simple; they show that on the space of solutions to Einstein's equations, the mass function has a non-degenerate critical point at flat, or Minkowski, space.

An important feature of the work of Brill and Deser is that the infinite dimensional techniques employed are natural, useful and indispensible.

Linearization Stability.

Another fundamental problem in general relativity which has been solved using techniques from global analysis is that of <u>linearization stability</u>. This problem may be explained as follows. Suppose we have a given spacetime, for example the Schwarzchild metric, and then wish to consider a slightly perturbed situation; for instance the introduction of a slight irregularity or a small planet. To consider such situations directly is not easy because of the non-linear nature of Einstein's equations. Instead, it is common to linearize the equations, solve these linearized equations, and assert that the solution is an approximation to the "true" solution of the non-linear equations.

To motivate linearization, we consider the following standard perturbation argument.

Suppose $\text{Ric}(g_0) = 0$ and we seek to solve for g near g_0. Write $g(\lambda)$ for a parameter λ and expand:

$$g(\lambda) = g_0 + \lambda h + \lambda^2 k + \ldots .$$

The approximation to first order is $g + \lambda h$. Now $h = \frac{dg}{d\lambda}\big|_{\lambda=0}$. If $\text{Ric}(g(\lambda)) = 0$ we find that for h,

$$D\,\text{Ric}(g_0) \cdot h = 0 . \qquad (2)$$

These are the linearized field equations (they are written out below).

It is perhaps surprizing that the implicit assumption -- that the solution of the linear equations approximates the solution of the full equations -- is not always valid. Such a possibility was indicated by Brill-Deser [2], and has been established rigorously by the authors[*] in the case the universe is "toroidal"; i.e. $T^3 \times R$ where T^3 denotes the flat 3-torus. If the above assumption on the given spacetime is valid, that spacetime is called <u>linearization stable</u>.

The theorem below shows that Brill's example is exceptional and that most spacetimes can be expected to be linearization stable. Although it would be unpleasant if this were not so, the example and the delicacy of the result show that caution is to be exercised when such sweeping assumptions are made.

<u>Theorem</u>. <u>Suppose that the</u> ("background") <u>spacetime with metric tensor</u> $g_{\alpha\beta}$ <u>satisfies the following conditions: there is a space like hypersurface</u> M <u>with induced metric</u> g <u>and second fundamental form</u> k <u>such that</u>

(R)
- (i) <u>there are no infinitesimal isometries</u> X <u>on both</u> g <u>and</u> k (<u>if</u> M <u>is not compact</u>, X <u>is required to vanish at infinity</u>)
- (ii) <u>if</u> k = 0 <u>and</u> M <u>is compact then</u> g <u>is not flat</u>
- (iii) <u>if</u> k $\neq$ 0, tr(k) = <u>trace of</u> k <u>is constant on</u> M <u>is</u> M <u>is compact, and</u> tr(k) = 0 <u>if</u> M <u>is non-compact</u>
- (iv) <u>if</u> M <u>is non-compact</u>, g <u>is complete and in a suitable sense asymptotically Euclidean</u>.

---

[*]General isolation theorems along these lines are given in Fischer-Marsden [6].

Then near M , the spacetime metric $g_{\alpha\beta}$ is linearization stable.

Brill's example fits in because condition (ii) fails for $M = T^3$, the flat 3-torus.

The following corollary was obtained by Choquet-Bruhat and Deser [1] independently.

Corollary. Minkowski space is linearization stable.*

Although the proof is complicated in details, we can endeavor to give the main ideas here. Further details will be filled in below. It is a simple and elegant application of the theory of infinite dimensional manifolds.

In order to solve the Einstein equations, one can, as explained in lecture 9, regard them as evolution equations with g , k (as given in the statement of the theorem) as initial, or Cauchy, data. However we must remember that there are the non-linear constraints to be imposed; i.e.:

$$\left. \begin{array}{l} \mathcal{H}(g, \pi) = \frac{1}{2}((\mathrm{tr}\pi)^2 - \pi\cdot\pi) + R(g) = 0 \\ \\ \delta(g, \pi) = -\pi^{ij}{}_{|j} = 0 \end{array} \right\} \quad (3)$$

This defines a certain **non-linear** subset C of $T\mathcal{M}$, the space of all g's and k's on M . The principal method is the following: near those g , k for which the conditions of the theorem are satisfied,

---
* The linearized equations in this case are referred to as the "weak field approximation" and can be used to study gravitational waves.

the set C is a smooth infinite dimensional submanifold of the space $T\mathcal{M}$. The other points are singular.

The smoothness of the set C entails that tangent vectors to C are closely approximated by points in C itself (which would not be the case if C has corners or other singularities). This remark together with existence theorems for the Einstein equations then yields the desired result.

Fortunately, establishing the smoothness of C can be done by techniques which we have previously developed in lectures 1 and 3.

Some Proofs.

Now we shall fill in a few proofs of the above results. We begin by studying the constraint manifold.

Let $C_{\mathcal{H}} = \mathcal{H}^{-1}(0) = \{(g, \pi) | \mathcal{H}(g, \pi) \equiv 0\}$.

Theorem 1. Let $(g, \pi) \in C_{\mathcal{H}}$ satisfy condition (ii) of (R). Then in a neighborhood of (g, π), $C_{\mathcal{H}}$ is a smooth submanifold of $T\mathcal{M}$.

Proof. Consider $\mathcal{H} : T\mathcal{M} \to C^{\infty}$. We show that $D\mathcal{H}(g, \pi)$ is onto. It follows that $\mathcal{H}$ is a submersion at g, π so that $\mathcal{H}^{-1}(0)$ is then a smooth submanifold in a neighborhood of (g, π) (see lecture 1). From A. Lichnerowicz [2] we have the classical formula

$$DR(g) \cdot h = \Delta(\mathrm{tr}\, h) + \delta\delta h - h \cdot \mathrm{Ric}(g) \qquad (4)$$

and from this one finds

$$D\mathcal{H}(g, \pi) \cdot (h, \omega) = 2\{\tfrac{1}{2}(\mathrm{tr}\pi)\pi - \pi \times \pi\} \cdot h + \Delta(\mathrm{tr}h) + \delta\delta h - h \cdot \mathrm{Ric}(g)$$

$$+ 2\{\tfrac{1}{2}(\mathrm{tr}\pi)g^{-1} - \pi\} \cdot \omega \qquad (5)$$

where $\pi \times \pi = \pi^{ik}\pi_{kj}$. Using elliptic theory, it follows that $D\mathcal{H}(g, \pi)$ is surjective provided that its adjoint $D\mathcal{H}(g, \pi)^*$ is injective and has injective symbol. A straightforward computation shows that $D\mathcal{H}(g, \pi)^* : C^\infty \to S_2 \times S^2$ is given by

$$D\mathcal{H}(g, \pi)^* \cdot N = (2\{\tfrac{1}{2}(\mathrm{tr}\pi)\pi - \pi \times \pi\}^{-1}N + g\Delta N + \mathrm{Hess}\, N - N\, \mathrm{Ric}(g),$$

$$2\{\tfrac{1}{2}(\mathrm{tr}\pi)g^{-1} - \pi\}N) . \qquad (6)$$

<u>Note</u>. The symbol of $D\mathcal{H}(g, \pi)^*$ is $\sigma_\xi(s) = (-g\|\xi\|^2 + (\xi \times \xi)^{-1}, 2(\tfrac{1}{2}(\mathrm{tr}\pi)g^{-1} - \pi))s$; $\xi \in T_xM$, $s \in S_2(M)_x$, which is always injective, so one has the L_2 orthogonal splitting $C^\infty = \ker D\mathcal{H}^* \oplus \mathrm{range}\,(D\mathcal{H})$; see lecture 3.

If $N \in \mathrm{kernel}\,(D\mathcal{H}(g, \pi)^*)$ then

$$g\Delta N + \mathrm{Hess}\, N - N\,\mathrm{Ric}(g) + 2(\tfrac{1}{2}(\mathrm{tr}\pi)\pi - \pi \times \pi)N = 0 \qquad (7)$$

and

$$2(\tfrac{1}{2}(\mathrm{tr}\pi)g^{-1} - \pi)N = 0 . \qquad (8)$$

Taking the trace of (7) gives $(\mathrm{tr}\pi) \cdot N = 0$, so (8) gives $N\pi = 0$. Thus, (7) gives

$$g\Delta N + \mathrm{Hess}\, N - N \cdot \mathrm{Ric}(g) = 0 \qquad (9)$$

whose trace gives

$$2\Delta N - N \cdot R(g) = 0. \tag{10}$$

Using $\mathcal{H}(g, \pi) = 0$, $N\pi = 0$, (10) becomes $\Delta N = 0$, so N is constant. If $\pi \neq 0$ this gives $N = 0$. If $\pi = 0$ then from (7), $N \text{ Ric}(g) = 0$ so as g is not flat in this case, $\text{Ric}(g) \neq 0$ as we are on a 3-manifold and so $N = 0$. □

In the non-compact case N constant would automatically force $N = 0$ by using suitable asymptotic conditions.

By the same methods one can prove the following theorem in geometry.

<u>Theorem 2.</u> <u>Let</u>[*] $\rho \in C^\infty$, $\rho \leq 0$, $\rho \neq 0$. <u>Then</u> $\mathfrak{m}_\rho = \{g \in \mathfrak{m} | R(g) = \rho\}$ <u>is a smooth submanifold of</u> $\mathfrak{m}$, <u>as is</u> $\mathfrak{m}_0' = \{g \in \mathfrak{m} | R(g) = 0$, g <u>not</u> <u>flat</u>$\}$.

This result enables one to handle the time symmetric case directly (i.e. $\pi = 0$ on M), if we restrict to deformations respecting the time symmetry.

If (ii) of (R) is not fulfilled, i.e. if $k = 0$ and if g is flat, then $D\mathcal{H}(g, \pi)$ is not a submersion. In fact the behaviour near these points is rather different. For example, using an idea of Brill-Deser [2] one can show that if g is flat, solutions of $R(g) = 0$ near g are obtainable from g by a coordinate transformation and the addition of a covariant constant $h \in S_2$, so in particular are flat.

---

[*] A refinement of the argument due to J.P. Bourguinon shows that we only need $\rho \neq 0$ and ρ not a positive constant. See Fischer-Marsden [6].

Next we investigate the divergence constraint. Let

$$C_\delta \subset T\mathfrak{M}, \quad C_\delta = \{(g, \pi) \mid \delta\pi = -\pi^{ij}{}_{|j} = 0\}.$$

<u>Theorem 3.</u> <u>Let</u> $(g, \pi) \in C_\delta$ <u>satisfy the following condition:</u>
(i)': <u>for a vector field</u> X, $L_X g = 0$ <u>and</u> $L_X \pi = 0$ <u>implies</u> $X = 0$.
<u>Then</u> C_δ <u>is a smooth submanifold in a neighborhood of</u> (g, π).

<u>Proof.</u> One computes that $D\delta(g, \pi) : S_2 \times S^2 \to \mathfrak{X}$ is given by:

$$D\delta(g, \pi) \cdot (h, \omega) = \delta\omega + \tfrac{1}{2}\pi^{jk} h_{jk}{}^{|i} - \pi^{jk} h^i{}_{j|k} - \tfrac{1}{2}\pi^{ij}(\mathrm{tr}\, h)_{|j} \quad (11)$$

and its adjoint $D\delta(g, \pi)^* : \mathfrak{X} \to S_2 \times S_2$ is:

$$D\delta(g, \pi)^* \cdot X = (\{-\tfrac{1}{2} L_X \pi + \tfrac{1}{2}(\delta X)\pi - (X \otimes \delta\pi + \delta\pi \otimes X)\}^b$$
$$+ \tfrac{1}{4}(L_X g \cdot \pi) g - \tfrac{1}{2}(X \cdot \delta\pi) g, \tfrac{1}{2} L_X g) \quad (12)$$

(^{b} denotes that the indices are lowered). The symbol of $D\delta^*$ is again injective, so it suffices to show $D\delta(g, \pi)^* X = 0$ implies $X = 0$. Since $\delta\pi = 0$, the condition $D\delta(g, \pi)^* X = 0$ reads:

$$(-\tfrac{1}{2} L_X \pi + \tfrac{1}{2}(\delta X)\pi)^b + \tfrac{1}{4}(L_X g \cdot \pi) g = 0 \quad (13)$$

and

$$L_X g = 0. \quad (14)$$

From (14), $\delta X = 0$ so (13) gives $L_X \pi = 0$. Thus $X = 0$ by (i)'. □

The regular points satisfying (i)' are just those (g,π) having discrete isotropy group under the action of the diffeomorphism group. It is known that almost every g has no isometries (Ebin [1]).

To show that $C = C_{\mathcal{H}} \cap C_\delta$ is a submanifold, we need additional restrictions because there may be points at which the intersection is not transversal. This extra condition is (iii) of the conditions (R) on p. 232.

Theorem 4. *Let* $(g, \pi) \in C$ *satisfy conditions* (R). *Then in a neighborhood of* (g, π), C *is a smooth submanifold.*

Proof. Consider $\Phi = (\mathcal{H}, \delta) : \mathbb{T}\mathfrak{m} \to C^\infty \times \mathfrak{X}$. We want to show $D\Phi(g, \pi)$ is surjective; this will show that $\Phi^{-1}(\{0\} \times \{0\}) \subset \mathbb{T}\mathfrak{m}$ is a submanifold and will give the result. We know $D\Phi$ from (5) and (11). The adjoint map is given as follows:

$$D\Phi(g, \pi)^* : C^\infty \times \mathfrak{X} \to S_2 \times S^2 : (N, X) \to \Big((\Delta N)g + \text{Hess } N - N \text{ Ric}(g)$$
$$+ \{2(\tfrac{1}{2}(\text{tr}\pi)\pi - \pi \times \pi)N - \tfrac{1}{2} L_X \pi + \tfrac{1}{2}(\delta X)\pi - \tfrac{1}{2}(X \otimes \delta\pi + \delta\pi \otimes X)\}^b \quad (15)$$
$$+ \tfrac{1}{4}(L_X g \cdot \pi)g - (X \cdot \delta\pi)g \;,\; 2\{\tfrac{1}{2}(\text{tr}\pi) g^{-1} - \pi\}N + (\tfrac{1}{2}(L_X g))\Big).$$

The symbol of this map, $\sigma_\xi(D\Phi(g, \pi)^*)$, $\xi \in T_x M$ may be shown to be injective. (Indeed, one must show $\xi \neq 0$, and $\sigma_\xi(N, X) = \Big(g\|\xi\|^2 N$
$+ \xi \otimes \xi N + \tfrac{1}{2}\pi^{ik}\xi_k X^j + \tfrac{1}{2}\pi^{jk}\xi_k X^i - \tfrac{1}{2}\pi_{ij}\xi_k X^k + \tfrac{1}{4} g_{ij}\pi^{k\ell}(\xi_k X_\ell + \xi_\ell X_k)$

$\tfrac{1}{2}(\xi_i X_j + \xi_j X_i)\Big) = 0$ implies $X, N = 0$. This may be shown by contracting the second component with $g, \xi^i \xi^j$ and $\xi^i X^j$ to give $X = 0$ Then the first component gives $N = 0$.) Thus it remains to show that $D\Phi(g, \pi)^*$ is injective. Let $N, X \in \ker(D\Phi(g, \pi)^*)$. Since $\delta\pi = 0$, (15) gives

$$(\Delta N)g + \text{Hess } N - N \text{ Ric}(g) + 2\{\tfrac{1}{2}(\text{tr}\pi)\pi \times \pi\}N$$
$$- \tfrac{1}{2} L_X \pi + \tfrac{1}{2}(\delta X)\pi + \tfrac{1}{4}(L_X g \cdot \pi)g = 0 \tag{16}$$

and $\{\tfrac{1}{2}(\text{tr}\pi)g - \pi\}N + \tfrac{1}{2} L_X g = 0$ (17)

Taking the trace and using $\mathcal{H}(g, \pi) = 0$ gives:

$$2\Delta N + 2\{\pi \cdot \pi - \tfrac{1}{4}(\text{tr}\pi)^2\}N - \tfrac{1}{2}(d\text{tr}\pi) \cdot X = 0 \tag{18}$$

and $2\{\tfrac{1}{2}(\text{tr}\pi)g - \pi\}N + \tfrac{1}{2} L_X g = 0$ (19)

If $\pi = 0$ then (18) gives $N = $ constant and (16) gives $N\text{Ric}(g) = 0$ so $N = 0$, as $\text{Ric}(g) \neq 0$ in this case. By (19), and (i) of (R) we obtain $X = 0$.

If $\pi \neq 0$, $\text{tr}\pi = $ constant, so (18) gives

$$2\Delta N + 2\{\pi \cdot \pi - \tfrac{1}{4}(\text{tr}\pi)^2\}N = 2\Delta N + 2(\pi - \tfrac{1}{4}(\text{tr}\pi)g)^2 N = 0 \text{ , using } g \cdot g = 3.$$

Since $(\pi - \tfrac{1}{4}(\text{tr}\pi)g)^2 = (\pi - \tfrac{1}{4}(\text{tr}\pi)g) \cdot (\pi - \tfrac{1}{4}(\text{tr}\pi)g) > 0$, we conclude that $N = 0$. Then as before, $X = 0$. □

Integration of Infinitesimal Deformations of Ricci Flat Spacetimes

As explained previously, we can use Theorem 4 to prove the following result.

Theorem 5. Let $^{(4)}g$ be a Lorentz metric on V satisfying (1): $\text{Ric}(^{(4)}g) = 0$. Let $^{(4)}h$ satisfy the linearized equations (2); i.e.

$$\text{DRic}(^{(4)}g) \cdot {}^{(4)}h = \tfrac{1}{2}\left(\Delta_L {}^{(4)}h - \alpha_{(4)g}\delta({}^{(4)}h - \tfrac{1}{2}(\text{tr}\,{}^{(4)}h)g)\right) = 0 \tag{19}$$

where $\alpha_g(X) = L_X g$ and Δ_L is the Lichnerowicz d'Alembertian Lichnerowicz [21].

Let M be a compact oriented space-like hypersurface in V with induced metric g and second fundamental form k. Assume g, k satisfy the conditions (R).

Then there exists a $\delta > 0$ and a smooth curve $^{(4)}g(\lambda)$ of exact solutions of $\text{Ric}(^{(4)}g(\lambda)) = 0$ such that $^{(4)}g(0) = {}^{(4)}g$ and $\frac{d}{d\lambda}{}^{(4)}g(\lambda)\big|_{\lambda=0} = {}^{(4)}h$.

Proof. In Gaussian Normal coordinates in a neighborhood of M, $^{(4)}h$ induces a deformation h, ω by

$$h_{ij} = {}^{(4)}h_{ij}, \quad \omega = D(\pi^{ij}(g, \dot g)) \cdot (h, \dot h)$$

$$= -(h \times \pi + \pi \times w) + \tfrac{1}{2}(\dot h - g\,\text{tr}\,\dot h)^{-1} + \tfrac{1}{2}(h \cdot \dot g)g - (\text{tr}\,\dot g)h)^{-1}$$

where $\pi^{ij} = \tfrac{1}{2}(g^{ik}g^{j\ell}\dot g_{k\ell} - \tfrac{1}{2}(\text{tr}\,\dot g)g^{ij})$, $\dot g_{ij} = \partial\,{}^{(4)}g_{ij}/\partial t$ etc.

This induced deformation h, ω satisfies the linearized constraint equations: $D\mathcal{H}(g, \pi) \cdot (h, \omega) = 0$, $D\delta(g, \pi) \cdot (h, \omega) = 0$. In other words, $(h, w) \in T_{(g, \pi)}\mathcal{C}$.

Thus by Theorem 4, we can find a curve $(g(\lambda), \pi(\lambda)) \in C$ tangent to (h,ω) at (g, π). This gives us spacetimes $^{(4)}g(\lambda)$ defined on a neighborhood of M by the existence theory.

The only thing is that $\partial^4 g(\lambda)/\partial \lambda$ at $\lambda = 0$ may not match $^{(4)}h$ in the $0,0$ or $0i$ components. But this can be achieved by using a suitable lapse and shift; i.e. a suitable space time coordinate transformation. See lecture 9. □

APPENDIX

ON THE CORRESPONDENCE PRINCIPLE IN QUANTUM MECHANICS

Introduction

The problem we are concerned with is showing, in a suitable sense, that the solutions of the Schrodinger equation converge to solutions of Hamilton's equation as $\hbar \to 0$.

The formal resolution of this problem has been known since 1930 (cf. Frenkel [1]) via Hamilton-Jacobi theory. However, these formal resolutions show that the equations converge and not that the solutions of the equations converge. The last step is not simple because it is, in essence, a singular perturbation problem for non-linear partial differential equations.

A more complete treatment with refined formulas has been given by Maslov [1] (cf. Arnold [2]). Unfortunately this approach is quite complicated.

The aim of this appendix is to outline a simple proof of the convergence of the solutions as $\hbar \to 0$ by using the Hamilton-Jacobi theory, the "hydrodynamic" formulation of quantum mechanics and recent theorems on vanishing viscosity in hydrodynamics (see lectures 4 and 5 and Ebin-Marsden [1], Marsden [5]).

We shall be working on $\mathbb{R}^3$ to simplify the exposition. It can be generalized to "asymptotically flat" or compact Riemannian manifolds and multiparticle systems as well.

Hamilton-Jacobi Equation; The Classical Equations

The following picture of classical mechanics is standard (cf. Synge and Griffiths [1]).

Consider the Hamilton-Jacobi equation for $S(x,t)$, $x \in \mathbb{R}^3$, $t \in \mathbb{R}$,

$$\frac{\partial S}{\partial t} + H(x, \nabla S) = 0 \qquad (1)$$

where

$$H(x,p) = \frac{|p|^2}{2m} + U(x) .$$

The solution of this equation is related to the classical trajectories as follows: let S_o be S at $t = 0$, let $x_o \in \mathbb{R}^3$ and $p_o = \nabla S_o(x_o)$. The classical trajectory starting at x_o, p_o is, say $x(t)$, $p(t)$ so $\dot{x} = p/m$, $\dot{p} = -\nabla U$. Then $\nabla S(t, x(t)) = p(t)$. Thus the "waves of action" defined by $S = $ constant sweep out the classical trajectories.

Consider a classical statistical state represented by an initial probability function ρ on $\mathbb{R}^3$. This would then evolve by letting it flow along the trajectories; i.e., by the equation of continuity:

$$\frac{\partial \rho}{\partial t} + \text{div}\, [\rho \nabla S/m] = 0 \qquad (2)$$

Thus we may regard (1) and (2) as a way of propagating a probability density ρ on configuration space with initial momenta ∇S_o via classical trajectories.

The Quantum Equations

Consider a solution $\psi(t,x)$ of Schrodingers equation:

$$\frac{\hbar}{i} \frac{\partial \psi}{\partial t} = \frac{\hbar^2}{2m} \Delta \psi + U(x)\psi \tag{3}$$

write $\psi = \sqrt{\rho}\, e^{iS/\hbar}$ so that ρ is a probability measure on $\mathbb{R}^3$. Writing out (3) gives

$$\frac{-\hbar^2}{2m} \Delta \sqrt{\rho} + \frac{\partial S}{\partial t} + \frac{1}{2m}(\nabla S)^2 + U = 0$$

$$\frac{\partial \rho}{\partial t} + \text{div}\,[\rho\, \nabla S/m] = 0 \;. \tag{4}$$

Clearly as $\hbar \to 0$, (4) reduces to (1) and (2), so the solutions and in particular $\rho_\hbar$ ought to converge as $\hbar \to 0$ to the solutions of (1) and (2). However, the equations (4) are non-linear and $\hbar$ is involved in the highest order term, so such a conclusion is far from obvious.

The Resolution via Hydrodynamics

Theorem. <u>Fix an initial ρ and dS. Let $\rho_\hbar(t)$ denote the solution of (4) (i.e., of (3)) and $\rho(t)$ that of (1), (2). Assume, e.g., U is C with compact support. Then for each t, as $\hbar \to 0$, $\rho_\hbar(t) \to \rho(t)$ in each Sobolev space H^s, s > 5 and hence in the C^∞ topology; $0 \leq |t| \leq T$ for some $T > 0$.</u>

One proves this by rewriting (4) by taking the gradient of the first equation; this yields the hydrodynamic model:

$$\begin{cases} \frac{\hbar^2}{2m} \Delta(\nabla\sqrt{\rho}) + \frac{\partial}{\partial t}(\nabla S) + \frac{1}{2n}\nabla(\nabla S) \cdot \nabla S + \nabla U = 0 \\ \\ \frac{\partial \rho}{\partial t} + \text{div}(\rho \nabla S/m) = 0 \end{cases}$$

These equations are the same as the equations for compressible flow $v = \nabla S/m$ with the extra "forcing" term

$$\nabla U - \frac{\hbar^2}{2m} \Delta(\nabla\sqrt{\rho}) \quad \text{(or pressure} \quad p = U - \frac{\hbar^2}{2m}(\Delta\sqrt{\rho})).$$

While $-\frac{\hbar^2}{2m}\Delta(\Delta\sqrt{\rho})$ is not a viscosity term, it can be handled in the same way, as in [2], [5], [6]. One uses "Lagrangian" coordinates and Trotter product formulas to show that the error is $\|\sqrt{\rho_\hbar} - \sqrt{\rho}\| = O(\hbar^2)$ in H^2 norm. In fact it is a general result that if a linear generator arising from a differential operator νT added to the hydrodynamic equations (with no boundaries present) then the solutions converge as $\nu \to 0$.

Remarks. 1. This technique only seems to work for sufficiently smooth U. It would be interesting and non-trivial to extend this to the hydrogen atom.

2. There are topological obstructions to obtaining S for all t by this method since the trajectories in configuration space can cross; i.e., a "shock" can develop. The correction required is the Maslov index, and it allows the previous analysis to be extended globally in time.

BIBLIOGRAPHY

Abraham, R.
[1] "Lectures of Smale on Differential Topology." (mimeographed, Columbia, 1961).
[2] "Foundations of Mechanics," Benjamin, 1967.
[3] Piecewise Differentiable Manifolds and the Space-Time of General Relativity, Journ. Math. and Mech. 11 (1962) 553-592.
[4] Introduction to Morphology, Publications du departement de Mathematiques de Lyon, Vol. 4 (1972) 38-114.
[5] Hamiltonian Catastrophies, Publications du departement de Mathematiques de Lyon, Vol. 4 (1972) 1-37.
[6] "Lectures on Global Analysis", Mimeographed, Princeton (1968).

Abraham, R., and Robbin, J.
[1] "Transversal Mappings and Flows," W. A. Benjamin Inc., N. Y. 1967.

Adler, S., Bazin, R., and Schiffer, M.
[1] "Introduction to General Relativity," McGraw-Hill (1966).

Alexander, J. C., and Yorke, J. A.
[1] Global bifurcation of periodic orbits, (to appear).

Andrea, S. A.
[1] The Hodge Existence Theorem (I,II), Advances in Math. 6 (1971) 389.

Antman, P., and Keller, J.
[1] "Bifurcation theory and Nonlinear Eigenvalue Problems," Benjamin (1969).

Andronov, A., and Chaikin, C.
[1] "Theory of Oscillations" (S. Lefschetz ed.)., Princeton University Press 1949.

Antonelli, P. L., Burghelea, D., and Kahn, P. J.
[1] The Nonfinite type of some $\text{Diff}_0(M^n)$, Bull. Am. Math. Soc. 86 (1970) 1246-1250 and Topology 11 (1972) 1-50.

Arnold, V.
[1] Sur la geometrie differentielle des groupes de Lie de dimension infinie et ses applications a l'hydrodynamique des fluids parfaits, Ann. Inst. Grenoble 16 (1) (1966), 319-361.
[2] Characteristic Class Entering in Quantization Conditions, Funct. An. and Appl. 1 (1967) 1-13.

Arnowitt, R., Deser, S., and Misner, C.W.
 [1] The Dynamics of General Relativity, in "Gravitation; an introduction to current research," ed. L. Witten, Wiley, New York, 1962.
 [2] Dynamical Structure and Definition of Energy in General Relativity, Phys, Rev. 116 (1959) 1322-1330.

Avez, A.
 [1] Essais de geometric riemannienne hyperbolique global- applications a la relativite general. Ann. Inst. Four. Grenoble 13 (1963) 105-190.
 [2] Le Probleme des Conditions Initiales, 163-167 in "Fluides et champs gravitationnels en relativite generale." Colloques internationaux du C.N.R.S. no. 170, Paris (1967).

Ball, J.M.
 [1] Continuity Properties of Nonlinear Semigroups (to appear).

Bardos, C., and Tartar, L.
 [1] Sur l'unicite retrograde des equations parabolique et quelque questions voisines. Arch. Rat. Mech. and Anal. 50 (1973) 10-25.

Bass, J.
 [1] Les fonctions pseudo-aleatoire, Memorial des Sciences Mathematique, Fascicule CLIII, Gauthier-Villars, Paris (1962).
 [2] Fonctions stationnaires. Fonctions de Correlation. Application a la representation spatio-temporelle de la turbulence, Ann. Inst. H. Poincare, 5 (1969) 135-193.

Batchelor, G.K.
 [1] "The Theory of Homogeneous Turbulence," Cambridge University Press (1953).

Bell, J.S.
 [1] On the Problem of Hidden Variables in Quantum Mechanics, Rev. Mod. Phys. 38 (1966), 447-552.

Berger, M.
 [1] Quelques formules de variation pour une structure riemannienne, Ann. Scient. Ec. Norm. Sup. $\underline{3}$ (1970) 285-294.
 [2] Sur les varietes d'Einstein compactes (in Comptes Rendus III[e] reunion math. expression latine, Namur 1965).

Berger, M., and Ebin, D.
[1] Some Decompositions of the Space of Symmetric Tensors on a Riemannian Manifold, J. Diff. Geom. 3 (1969) 379-392.

Birkhoff, G.
[1] "Hydrodynamics; a Study in Logic, Fact and Similitude," Princeton University Press (1950).

Bishop, R., and Crittenden, R.
[1] "Geometry of Manifolds," Academic Press, N.Y. 1964.

Bohm, D., and Bub, J.
[1] A Proposed Solution of the measurement problem in quantum mechanics by a Hidden Variable Theory, Rev. Mod. Phys. 38 (1966), 453-469, 470-475.

Bub, J.
[1] On the Completeness of Quantum Mechanics in "Contemporary Research in the Foundations and Philosophy of Quantum Theory", ed. C.A. Hooker, D. Reidel Pub. Co. Boston, (1973).

Bowen, R.
[1] Markov Partitions for Axiom A diffeomorphisms, Am. Journ. Math. 92 (1970) 725-747, 907-918.
[2] Periodic Points and Measures for Axiom A Diffeomorphisms, Trans. Am. Math. Soc. 154 (1971) 377-397.

Bourbaki, N.
[1] Varietes differentiables et analytiques - Fascicule des resultats Hermann, Paris (1969).

Bourguignon, J.P., and Brezis, H.
[1] Remarks on the Euler Equations (preprint).

Brezis, H.
[1] On a Characterization of Flow Invariant Sets, Comm. Pure and Appl. Math. 23 (1970) 261-263.

Brezis H., and Pazy, A.
[1] Semi-groups of nonlinear contractions on convex sets. J. Funct. An. 6 (1970) 232-281, 9 (1972) 63-74.

Brill, D., and Deser, S.
[1] Variational Methods and Positive Energy in Relativity, Ann. Phys. 50 (1968) 548-570.
[2] Instability of Closed Spaces in General Relativity, Comm. Math. Phys. 32 (1973) 291-304.

Browder, F.
[1] Existence theorems for non-linear partial differential equations, Proc. Symp. Pure Math. Am. Math. Soc. XVI (1970) 1-60.

Bruslinskaya, N.N.
[1] The origin of cells and rings at near-critical Reynolds numbers, Uspekhi Mat. Nauk. 20 (1965) 259-260.
[2] The Behavior of Solutions of the Equations of Hydrodynamics when the Reynolds number passes through a critical value, Doklady 6 (4) (1965) 724-728.
[3] Qualitative Integration of a system of n differential Equations in a region containing a singular Point and a Limit cycle, Dokl. Akad. Nauk, SSR 139 (1961) 9-12.

Cantor, M.
[1] The group $\mathcal{L}_I^s (R^n)$ (preprint).
[2] Global analysis over non-compact spaces, Thesis, Berkeley 1973.
[3] Sobolev Inequalities for Riemannian Bundles, Proc. Symp. Pure Math, AMS (to appear).

Carroll, R.
[1] "Abstract Methods in Partial Differential Equations" Harper and Row (1969).

Cartan, E.
[1] Sur les équationes de la gravitation d'Einstein, J. Math. Pures et App. 1 (1922) 141-203.

Cerf, J.
[1] Topologie de certains Espaces de plongements, Bull. Soc. Math. France 89 (1961) 227-380.
[2] "Sur les diffeomorphismes de la sphere de dimension trois (Γ_4 = 0)." Springer Lecture Notes #53, 1968.

Chadam, J.M.
[1] Asymptotics for $\Box u = m^2 u + G$, I Global Existence and Decay, Bull. Am. Math. Soc. 76 (1970) 1032-5.

Chafee, N.
[1] The Bifurcation of One or More Closed Orbits from an Equilibrium Point of an Autonomous Differential System, Journ. Diff. Equations 4 (1968) 661-679.

Chandresekar, S.
[1] "Hydrodynamic and Hydromagnetic Stability," Oxford University Press (1961)

Chernoff, P.
[1] Note on Product Formulas for Operator Semi-Groups, J. Funct. An. 2 (1968) 238-242.
[2] Essential Self-Adjointness of Powers of Generators of Hyperbolic Operators, J. Funct. An. 12 (1973) 401-414.

Chernoff, P., and Marsden, J.
[1] "Hamiltonian Systems and Quantum Mechanics," (in preparation).
[2] On Continuity and Smoothness of Group Actions, Bull. Am. Math. Soc. 76 (1970) 1044.
[3] Some Remarks on Hamiltonian Systems and Quantum Mechanics (to appear).

Chichilnisky, G.
[1] Group Actions on Spin Manifolds, Trans. Am. Math. Soc. 172 (1972) 307-315.

Chillingworth, D., (ed).
[1] "Proceedings of the Symposium on differential equations and dynamical systems." Springer lecture Notes in Mathematics 206 (1971).

Choquet, G.
[1] "Lectures on Analysis" (3 vols.) W. A. Benjamin (1969).

Choquet-Bruhat, Y.
[1] Espaces-temps einsteiniens generaux chocs gravitationels, Ann. Inst. Henri Poincare, 8 (1968) 327-338.
[2] Solutions C^∞ d'equations hyperboliques non lineares, C.R. Acad. Sc. Paris, 272 (1971) 386-388.
[3] New Elliptic System and Global Solutions for the Constraints Equations in General Relativity, Comm. Math. Phys. 21 (1971) 211-218.

Choquet-Bruhat, Y., and Deser, S.
[1] Stabilite initiale de L,espace temps de Minkowski, C.R. Acad, Sc. Paris 275 (1972) 1019-1027.
[2] On the Stability of flat space (to appear).

Choquet-Bruhat, Y., and Geroch, R.
[1] Global Aspects of the Cauchy Problem in General Relativity, Comm. Math. Phys. 14 (1969) 329-335.

Chorin, A. J.
[1] On the Convergence of Discrete Approximations to the Navier-Stokes Equations, Math. of Comp. 23 (1969) 341-353.
[2] Numerical Study of Slightly Viscous Flow, Journ. Fluid Mech. 57 (1973) 785-796.
[3] Representations of Random Flow (to appear)

Clauser, J.F., Horn, M.A., Shimony, A., and Holt, R.A.
 [1] Proposed Experiment to test local Hidden Variable
     Theories, Phys. Rev. Letters 23 (1969) 880-884.

Coddington, E., and Levinson, N.
 [1] "Theory of Ordinary Differential Equations," McGraw-
     Hill, N.Y. (1955).

Coles, D.
 [1] Transition in Circular Couette Flow, Journ. Fluid
     Mech. 21 (1965) 385-425.

Cook, J.M.
 [1] Complex Hilbertion Structures on Stable Linear Dynamical
     Systems. J. Math. and Mech. 16 (1966) 339-349.

Courant, R., and Hilbert, D.
 [1] "Methods of Mathematical Physics," vol. I, II, Inter-
     science, New York, 1953, 1962.

Crandall, M. and Rabinowitz, P.
 [1] Bifurcation from Simple Eigenvalues, Journ. Funct.
     An. 8 (1971) 321-340, Arch. Rct. Mech. An 52 (1974) 161-180.

Cronin, J.
 [1] One-Sided Bifurcation Points, J. Diff. Eq ns. 9
     (1971) 1-12.

DeBroglie, L.
 [1] "The Current Interpretation of Wave Mechanics"
     Elsevier (1964).

Deser, S.
 [1] Covariant Decomposition of Symmetric Tensors and the
     Gravitational Cauchy Problem, Ann. Inst. H. Poincare VII
     (1967) 149-188.

DeWitt, B.
 [1] Quantum Theory of Gravity, I The Canonical Theory,
     Phys. Rev. 160 (1113-1148), 1967.
 [2] Space time as a Sheaf of Geodesics in Superspace, in
     "Relativity", ed. M. Carmeli, S. Fickler, and
     L. Witten, Plenum Press (1970).

Dieudonné, J.A.
 [1] "Foundations of Modern Analysis," Academic Press,
     New York, 1960.
 [2] "Treatise on Analysis," Vol. III. Academic Press,
     New York (1972).

Dionne, P.
[1] Sur les probleme de Cauchy bien poses. J. Anal. Math. Jerusalem 10 (1962/3) 1-90.

Dirac, P.A.M.
[1] Generalized Hamiltonian Dynamics, Can. Journ. Math. 2 (1950) 129-148.

Dixmier, J.
[1] "Les C^*-algebras et leurs representations", Gauthier-Villars (1964).

Duff, G.F.D.
[1] Differential forms in manifolds with boundary, Ann. of Math. 56 (1952), 115-127.
[2] On Turbulent Solutions of the Navier-Stokes Equations (unpublished).

Duff, G., and Spencer, D.
[1] Harmonic tensors on Riemannian manifolds with boundary, Ann. of Math. 56 (1952), 128-156.

Duff, G., and Naylor, D.
[1] "Partial Differential Equations of Applied Mathematics," Wiley, N.Y. (1969).

Dunford, N. and Schwartz, J.
[1] "Linear Operators," Vol. I, Interscience (1958).

Eardley, D., Liang, E., and Sachs, R.
[1] Velocity dominated singularities in irrotational dust cosmologies, Journ. Math. Phys. 13 (1972) 99-107.

Earle, C.J. and Eells, J.
[1] A fiber bundle description of Teichmuller theory, J. Diff. Geom. 3 (1969) p. 19-43 (also Bull. Am. Math. Soc. 73 (1967) 557-559).

Earle, C.J., and Schatz, A.
[1] Teichmuller theory for surfaces with boundary, J. Diff. Geom. 4 (1970) 169-186.

Ebin, D.G.
[1] The manifold of Riemannian metrics, in Proc. Symp. Pure Math. xv, Amer. Math. Soc. 1970, 11-40 and Bull. Am. Math. Soc. 74 (1968) 1002-1004.
[2] On Completeness of Hamiltonian Vector Fields, Proc. Am. Math. Soc. 26 (1970) 632-634.
[3] Espace des metriques riemanniennes et mouvement des fluides via les varietes d'applications, Lecture notes, Ecole Polytechnique, et Université de Paris VII(1972).
[4] Viscous fluids in a domain with frictionless boundary, J. Funct. An. (to appear).

Ebin. D.G., and Marsden, J.
[1] Groups of diffeomorphisms and the motion of an incompressible fluid, Ann. of Math. 92 (1970), 102-163. (See also Bull. Am. Math. Soc. 75 (1969) 962-967.)

Eells, J.
[1] On the geometry of function spaces, in Symposium de Topologia Algebrica, Mexico (1958) 303-307.
[2] A setting for global analysis, Bull. Amer. Math. Soc. 72 (1966) 751-807.

Eells, J., and Sampson, J.
[1] Harmonic Maps of Riemannian Manifolds, Am. J. Math 86 (1964) 109-160.

Einstein, A., Podolsky, B., and Rosen, N.
[1] Can Quantum Mechanical Description of Physical Reality be considered complete? Phys. Rev. 47 (1935)777-780.

Eliasson, H.
[1] Geometry of Manifolds of Maps. J. Diff. Geom. 1 (1967), 169-194.

Elhadad, J.
[1] Sur l'interpretation en geometrie symplectique des etats quantiques de l'atome d'hydrogene, Symposia Mathematica (to appear).

Epstein, D.B.A.
[1] The simplicity of certain groups of homeomorphisms, Compositio Math. 22 (1970) 163-173.

Fabes, E. B., Jones, B.F., and Rivierc, N.M.
[1] The initial value problem for the Navier-Stokes equations with data in L^P, Arch. Rat. Mech. An. 45 (1972) 222-240.

Faddeev, L.D.
[1] Symplectic structure and quantization of the Einstein gravitation theory, Actes du Congres Intern. Math. 3 (1970) 35-40.

Feynman, R. P., Leighton, R.B., and Sands, M.
[1] "The Feynman Lectures on Physics (II)," Addison-Wesley Co. Reading, Mass. (1964).

Finn, R.,
[1] On Steady State Solutions of the Navier-Stokes Equations, Arch. Rat. Mech. 3 (1959) 381-396, Acta Math. 105 (1961) 197-244.
[2] Stationary Solutions of the Navier-Stokes Equations, Proc. Symposia Appl. Math. 17 (1965) 121-153.

Finn, R., and Smith, D.R.
[1] On the Stationary Solutions of the Navier-Stokes Equations in Two Dimensions, Arch. Rat. Mech. Anal. 25 (1967) 26-39.

Fischer, A.
[1] The Theory of Superspace, in "Relativity," ed. M. Carmeli, S. Fickler and L. Witten, Plenum Press (1970).

Fischer, A., and Marsden, J.
[1] The Einstein Equations of Evolution - A Geometric Approach, J. Math. Phys. 13 (1972) 546-568.
[2] The Einstein Evolution Equations as a First Order Symmetric Hyberbolic Quasi-Linear System I, Comm. Math. Phys. 28 (1972) 1-38 II (in preparation).
[3] General Relativity, Partial Differential Equations and Dynamical Systems. Proc. Symp. Pure Math, AMS 23 (1973) 309-328.
[4] The Existence of Complete Asymptotically Flat Space-times, Publ. Dept. Math. Univ. Lyon. Vol. 4, Fasc. 2 (1972) 182-193.
[5] Linearization Stability of the Einstein Equations, Bull. Am. Math. Soc. 79 (1973) 995-1001.
[6] Deformations of non-linear partial differential equations, Proc. Symp. Pure Math (to appear).
[7] New Theoretical Techniques in the Study of Gravity, Gen. Rel. and Grav. 4 (1973) 309-318.
[8] Global Analysis an General Relativity, Gen. Rel. and Grav. 5 (1974) 89-93.

Fischer, A. and Wolf, J.
[1] The Calabi Construction for compact Ricci-flat riemannian manifolds, Bull. Am. Math. Soc. (to appear).

Flanders, H.
[1] "Differential Forms," Academic Press, New York (1963).

Foias, C.
[1] Une remarque sur l'unicite des solutions des equations de Navier-Stokes en dimension n. Bull. Soc. Math. France 89 (1961) 1-8.

Foias, C., and Prodi, G.
[1] Sur le comportement global des equations de Navier-Stokes en dimension 2. Rend. Sem. Mat. Padova XXXIX (1967) 1-34.

Foures-Bruhat, Y.
[1] Theoreme d'existence pour certains systems d'equations aux derivees partielles non lineaires, Acta. Math. 88 (141-225), 1952.

Freedman, S.J. and Clauser, J.F.
[1] Experimental Test of Local Hidden-Variable Theories, Phys. Rev. Letters, 28 (1972) 938-941.

Frenkel, J.
[1] "Wave Mechanics: Advanced General Theory," Oxford 1934.

Friedrichs, K.O.
[1] Symmetric hyperbolic linear differential equations, Comm. Pure and Appl. Math., 7 (1954) 345-392.
[2] "Special topics in fluid dynamics," Gordon and Breach, N.Y. 1966.

Friefeld, C.
[1] One Parameter Subgroups Do Not Fill a Neighborhood of the Identity in an Infinite Dimensional Lie (Pseudo) Group, in "Battelle Rencontres," ed. C.M. DeWitt and J.A. Wheeler, W.A. Benjamin (1968).

Fujita, H., and Kato, T.
[1] On the Navier-Stokes Initial Value Problem, I. Arch. Rat. Mech. Anal. 16 (1964) 269-315.

Gerlach, R.
[1] The Derivation of the Ten Einstein Field Equations From the Semiclassical Approximation to Quantum Geometrodynamics, Phys. Rev. 117 (1) (1969) 1929-1941.

Geroch, R.
[1] What is a Singularity in General Relativity? Ann. of Phys. 48 (1968) 526-540.
[2] Some recent work on global properties of spacetimes, Actes congres Intern. Math. 3 (1970) 41-46.

Golovkin, K.
[1] About Vanishing Viscosity in the Cauchy Problem for the Equations of Fluid Mechanics, Memoire of Steklov Inst. XCII, Moscow (1966).

Gordon, W.
[1] The Riemannian Structure of Certain Function Space Manifolds, J. Diff. Geom. 4 (1970), 499-508.
[2] On the completeness of Hamiltonian vector fields. Proc. Amer. Math. Soc. 26 (1970) 329-331.

Graff, D.
[1] Thesis, Princeton University (1971).

Gromov, M.L.
[1] Smoothing and Inversion of Differential Operators, Mat. Sbornik 88 (130) (1972) 381-435.

Gross, L.
[1] The Cauchy Problem for the coupled Maxwell and Dirac equations, Comm, in Pure and Appl. Math. 19 (1966)1-15.

Guynter, N.
[1] On the Basic Problem of Hydrodynamics, Trudy Fiz-Mat. Inst. Steklov, 2 (1927) 1-168. See also Isv. Akad. Nauk (1926) 1326, 1503 (1927) 621, 735, 1139, and (1928) 9.

Hall, W.S.
[1] The Bifurcation of Solutions in Banach Spaces, Trans. Am. Math. Soc. 161 (1971) 207-218.

Hartman, P.
[1] "Ordinary Differential Equations," Wiley, N.Y. (1964).
[2] The Swirling Flow Problem in Boundary Layer Theory, Arch. Rat. Mech. Anal. 42 (1971) 137-156.

Hawking, S.W. and Ellis, G.F.R
[1] "The large scale structure of spacetime," Cambridge (1973).

Hermann, R.
[1] "Differential Geometry and the Calculus of Variations," Academic Press, N.Y. (1968).
[2] "Lectures on Mathematical Physics, (I,II)", W.A. Benjamin, N.Y. (1971).
[3] Geodesics and classical mechanics on Lie groups, J. Math. Phys. 13 (1972) 460-464.

Herman, M.R.
[1] Simplicite du groupe de diffeomorphismes de class C^∞, isotopes a l'identite, du tore de dimension n, C.R. Acad. Sc., Paris t. 273 (1971) 232-234.

Herman, M., et Sergeraert, F.
[1] Sur un theoreme d'Arnold et Kolmogorov, C.R. Acad. Sc. Paris t. 273 (1971) 409-411.

Heywood J.
[1] On Stationary Solutions of the Navier-Stokes Equations as Limits of Nonstationary Solutions, Arch. Rat. Mech. and Anal. 37, No. 1 (1970), 48-60. See also Acta Math. 129 (1972) 11-34.

Hirsch, M., and Pugh, C.
[1] Stable Manifolds and Huperbolic Sets, Proc. Symp. Pure Math. XIV, Am. Math. Soc. (1970) 133-163.

Hirsch, M., Pugh, C., and Schub, M.
[1] Invariant Manifolds, Bull. Am. Math. Soc. 76 (1970) 1015-1019 and (in preparation).

Hodge, V.W.D.
 [1] "Theory and Applications of Harmonic Integrals,"
    Sec. Ed. Cambridge, 1952.

Holt, M. (ed).
 [1] "Proceedings of the Second International Conference
    on Numerical Methods in Fluid Mechanics," Springer
    Lecture Notes in Physics, Springer-Verlag (1971).

Hopf, E.
 [1] Abzweigung einer periodischen Losung von einer
    stationaren Losung eines Differential systems, Ber.
    Math-Phys. Sachsische akademie der Wissenschaften
    Leipzig 94 (1942) 1-22.
 [2] A Mathematical Example Displaying the Features of
    Turbulence, Comm. Pure Appl. Math. 1 (1948) 303-322.
 [3] Uber die Anfanswert-aufgabe fur die hydrodynamischen
    Grundgleichungen, Math. Nachr. 4 (1951) 213-231.
 [4] Repeated branching through loss of stability, An Example, Proc. Conf. on Diff. Equations, Univ. of
    Maryland (1955).
 [5] Remarks on the Functional-Analytic Approach to Turbulence, Proc. Symps. Appl. Math., XIII, Amer. Math.
    Soc. (1962) 157-163.
 [6] On the right weak solution of the Cauchy Problem for
    a quasilinear equations of first order, J. Math.
    Mech. 19 (1969/70) 483-487.

Hormander, L.
 [1] Fourier Integral Operators, Acta. Math., 127 (1971)
    79-183, 128 (1972) 183.

Iacob, A.
 [1] Invariant Manifolds in the Motion of a Rigid Body
    about a Fixed Point. Rev. Roum. Math. Pures et Appl.
    Rome XVI (1971) 149701521.
 [2] "Topological Methods in Mechanics" (in Roumanian)
    (1973).

Ikebe, T., and Kato, T.
 [1] Uniqueness of the self-adjoint extension of singular
    elliptic differential operators, Arch. Rat. Mech. 9
    (1962) 77-92.

Il'in, V.P. (ed).
 [1] "Boundary Value Problems of Mathematical Physics and
    Related Aspects of Function Theory" (I), Seminars in
    Mathematics (Leningrad) (5) Consultants Bureau,
    N.Y. (1969).

Iooss, G.
[1] Contribution a la theorie nonlineare de la stabilite des ecoulements laminaires, These, Faculte des Sciences, Pris VI (1971).
[2] Theorie non linearire de la stabilite des ecoulements laminaires dans le cas de << l'echange des stabilites >>, Arch. Rat. Mech. 40 (1971) 166-208.
[3] Existence de Stabilite de la Solution Periodique secondaire intervenant dans les problemes d'Evolution du Type Navier-Stokes, Arch. Rat. Mech. 49 (1972) 301-329.
[4] Bifurcation d'une solution T-periodique vers une solution nT-periodique, pour certains problèmes d'evolution du type Navier-Stokes, C.R. Acad. Sc. Paris 275 (1972) 935-938.
[5] Bifurcation et Stabilite, Lecture notes, Université Paris XI (1973)

Irwin, M.C.
[1] On the Stable Manifold Theorem, Bull. London Math. Soc. 2 (1970) 196-198.

Itaya, N.
[1] On the Cauchy Problem for the System of Fundamental Equations Describing the Movement of Compressible Viscous Fluid, Kodai Math. Sem. Rep. 23 (1971) 60-120.

Jauch, J.M.
[1] "Mathematical Foundations of Quantum Mechanics", Addison Wesely, (1968).

Joseph, D.D. and Sattinger, D.H.
[1] Bifurcating Time Periodic Solutions and their Stability Arch. Rat. Mech. and An. 45 (1972) 79-109.

Jost, R., and Zehnder, E.
[1] A generalization of the Hopf Bifurcation Theorem, Helv. Phys. Acta 45 (1972) 258-276.

Judovich, V.
[1] Non Stationary Flows of an Ideal Incompressible Fluid, Z. Vycisl. Mat. i, Fiz. 3 (1963), 1032-1066.
[2] Two-Dimensional Nonstationary Problem of the Flow of an Ideal Incompressible Fluid Through a Given Region, Mat. Sb. N.S. 64 (1964) 562-588.
[3] Example of the generation of a Secondary Stationary or Periodic Flow when there is a loss of Stability of the Laminar Flow of a Viscous Incompressible Fluid, Prikl, Math. Mek. 29 (1965) 453-467.

Judovich, V. (Cont'd.)
[4] Stability of convection flows, J. Appl. Math. Mech. 31 (1967) 294-303.
[5] The birth of proper oscillations in a fluid, Prikl. Math. Mek 35 (1971) 638-655.

Kahane, C.
[1] On the Spatial Analyticity of Solutions of the Navier-Stokes Equations, Arch. Rat. Mech. Anal. 33 (1969), 386-405.

Kaniel, S., and Shinbrot, M.
[1] A Reproductive Property of the Navier-Stokes Equations Arch. Rat. Mech. Anal. 24 (1967) 302-324.
[2] Smoothness of Weak Solutions of the Navier-Stokes Equations, Arch. Rat. Mech. Anal. 24 (1967) 302-323.

Kato, T.
[1] On Classical Solutions of the Two-Dimensional Non-Stationary Euler Equation, Arch. Rat. Mech. and Anal. 25 (3) (1967) 188-200.
[2] Nonstationary Flows of Viscous and Ideal Fluids in R^3. Journ. Funct. An. 9 (1972) 296-305.
[3] Linear Evolution Equations of "Hyperbolic" Type, Journ. Fac. Sci. Univ. of Tokyo, Sec. 1. XVII (1970) 241-258.
[4] Linear Evolution Equations of "Hyperbolic" type II (to appear).
[5] On the Initial Value Problem for Quasi-Linear Symmetric Hyperbolic Systems (to appear).
[6] "Perturbation Theory for Linear Operators," Springer (1966).

Kazdan, J., and Warner, F.
[1] Prescribing Curvatures, Proc. Sympos. Pure Math (AMS Summer Institute, Stanford 1973) (to appear) (see also Bull. Am. Math. Soc. 78 (1972) 570-574).

Keller, J.
[1] "Bifurcation theory for ordinary differential equations, in "Bifurcation Theory and Non-linear Eigenvalue Problems, ed J. Keller and S. Antman, Benjamin (1969).

Kirchgässner, K., and Kielhöfer, H
[1] Stability and bifurcation in fluid dynamics, Rocky Mountain J. of Math. 3 (1973) 275-318.

Kiselev, A.A., and Ladyzhenskaya, O.A.
[1] On Existence and Uniqueness of the Solution of the Non Stationary Problem for a Viscous Incompressible Fluid, Izvestiya Akad. Nauk, SSR 21 (1957) 655-680.

Knops, R.J., and Payne, L.E.
[1] On the Stability of Solutions of the Navier-Stokes Equations Backwards in Time, Arch. Rat. Mech. Anal. 29 (1968) 331-335.

Kochen, S., and Specker, E.P.
[1] The Problem of Hidden Variables in Quantum Mechanics, Journ. Math. and Mech. 17 (1967) 59-88.

Kodaira, K.
[1] Harmonic Fields in Riemannian Manifolds, Ann. of Math. 50 (1949), 587-665.

Kolmogorov, A.N.
[1] The Local Structure of Turbulence in Incompressible Viscous Fluid for Very Large Reynolds Numbers, C.R. Acad. Sci. USSR 30 (1941) 301.
[2] Dissipation of Energy in Locally Isotropic Turbulence, C.R. Acad. Sci. USSR 32 (1941) 16.

Kopell, N
[1] Commuting Diffeomorphisms, Proc. Symp. Pure Math. xiv (1970) 165-184.

Kopell, N., and Howard, L.
[1] Bifurcations under non-generic conditions, (to appear).

Kostant, B.
[1] Quantization and Unitary Representations, Lecture Notes in Math #170 Springer Verlag (1970).
[2] Orbits and Quantization Theory, Actes Conzris. Intern. Math. $\underline{2}$ (1970) 395-400.
[3] Symplectic Spinors, Symposia Mathematica (to appear).

Kraichnan, R.H.
[1] The Structure of Turbulence at Very High Reynolds Numbers, Journ. Fluids Mech. 5 (1959) 497-543.
[2] The Closure Problem of Turbulence Theory, Proc. Symp. Appl. Math. XIII, Am. Math. Soc. (1962) 199-225.
[3] Isotropic Turbulence and Inertial Range Structure, Phys. Fluids 9 (1966) 1728-1752.

Krikorian, N.
[1] Differentiable Structures on Function Spaces Trans. Am. Math. Soc. 171 (1972) 67-82.

Kundt, W.
[1] Global Theory of Spacetime, in Proc. 13th brennial seminar of Can. Math. Congress ed. J.R. Vanstone, Montreal (1972) 93-134.

Kunzle, H.
[1] Degenerate Lagrangian Systems, Ann. Inst. H. Poincare (A) XI (1969) 393-414.

Ladyzhenskaya, O.A.
[1] Solution "in the large" of Non Stationary Boundary Value Problem for the Navier-Stokes System with two Space Variables, Comm. Pure and Appl. Math. 12 (1959) 427-433.
[2] "The Mathematical Theory of Viscous Incompressible Flow" (2nd Edition), Gordon and Breach, N.Y. 1969.
[3] Example of Nonuniqueness in the Hopf Class of Weak Solutions for the Navier-Stokes Equations, Math. USSR-Izvestija 3 (1969) 229-236.
[4] (ed.) "Boundary Value Problems of Mathematical Physics and Related Aspects of Function Theory" (II,III), Seminars in Mathematics (Leingrad) (7,11) Consultants Bureau, N.Y. (1970).

Lanczos, C.
[1] "The Variational Principles of Mechanics" (2nd Edition), University of Toronto Press (1962).
[2] Einsteins Path from Special to General Relativity, in "General Relativity; papers in honour of J.L. Synge," L.O'Rafeartaigh (ed.), Oxford (1972).

Landau, L.D., and Lifshitz, E.M.
[1] "Fluid Mechanics," Addison-Wesley, Reading, Mass., 1959.

Lang, S.
[1] "Differential Manifolds," Addison-Wesley, Reading Mass, (1972).
[2] "Analysis II," Addison-Wesley, Reading Mass. 1969.

Lanford, O.E.
[1] Selected Topics in Functional Analysis, in "Statistical Mechanics and Quantum Field Theory ed. C. DeWitt and R. Stora, Gordon and Breach (1972).
[2] Bifurcation of periodic solutions into invariant tori: the work of Ruelle and Takens, in "Nonlinear Problems in the Physical Sciences and Biology," Springer Lecture Notes #322 (1973).

Lawruk, B., Sniatycki, J., and Tulczykjew, W.M.
[1] Special Symplectic Spaces (to appear).

Lax, P.
[1] Cauchy's Problem for Hyperbolic Equations and the Differentiability of Solutions of Elliptic Equations, Comm. Pure and Appl. Math 8 (1955) 615-633.

Leray, Jean
[1] Etude de diverses equations integrales non-lineares et de quelques problemes que pose l'hydrodynamique, Journ. Math. Pures, Appl. 12 (1933) 1-82.
[2] Essai sur les mouvements plans d'un liquide vesqueux que limitent des parois, J. de Math. 13 (1934). 331-418.
[3] Sur le mouvement d'un liquide visqueux emplissant l'espace, Acta Math 63 (1934) 193-248.
[4] Problemes non-lineaires, Ensign, Math. 35 (1936)139-151.
[5] Lectures on hyperbolic equations with variables coefficients, Inst. for Adv. Study, Princeton, 1952.

Leslie, J.
[1] On a differential structure for the group of diffeomorphisms, Topology 6 (1967), 263-271.
[2] Some Frobenius Theorems in Global Analysis,J. Diff. Geom. 2 (1968) 279-297.
[3] On two classes of Classical Subgroups of Diff(M), J. Diff. Geom. 5 (1971) 427-436.

Lichnerowicz, A.
[1] "Relativistic Hydrodynamics and Magnetohydrodynamics," Benjamin, 1967.
[2] Propagateurs et Commutateurs en relativite generale, Publ. Scientifiques, IHES #10. (1961) 293-344.

Lichtenstein, L.
[1] "Grundlagen Der Hydromechanik," Verlag Von Julius Springer, Berlin, 1929.
[2] Uber einige Existenzprobleme der Hydrodynamik homogener unzusammen-druckbarer, reibungsloser Flussigkeiter und die Helmholtzschen Wirbelsatze,Math. Z. 23 (1925), 89-154, 309-316; 26 (1927), 196-323; 28 (1928), 387-415, 725; 32 (1930), 608.

Lin, C.C.
[1] "The Theory of Hydrodynamic Stability," Cambridge University Press (1955).

Lions, J.L.
[1] Sur la regularite et l'unicite des solutions turbulentes des equations de Navier-Stokes. Rediconti Seminario Math. Univ. Padova 30 (1960) 16-23.
[2] Singular perturbations and singular layers in variational inequalities. in Zarantonello [1].
[3] "Perturbations Singulières dans les Problèmes aux Limites et en Contrôle Optimal, Springer Lecture Notes #323 (1973).
[4] "Quelques méthodes de résolution des problèmes aux limites non linéaires," Dunod, Gauthier Villars, 1969.

Lions, J. and Prodi, G.
[1] Un theoreme d'existence et unicite dan les equations de Navier-Stokes en dimension 2. C.R. Acad. Sci. Paris 248 (1959) 3519-21.

Lovelock, D., and Rund, H.
[1] Variational Principles in the General Theory of Relativity, Proc. 13th biennial seminar of Can. Math. Cong. ed. J.R. Vanstone Montreal (1972) 51-68.
[2] "Mathematical Aspects of Variational Principles in General Relativity", Lecture Notes, Univ. of Waterloo (1972).

Mackey, G.W.
[1] "Mathematical Foundations of Quantum Mechanics," Benjamin (1963).
[2] "Induced Representations of Groups and Quantum Mechanics" Benjamin (1969).

Maclane, S.
[1] Hamiltonian Mechanics and Geometry, Am. Math. Monthly, 77 (1970) 570-586.

Markus, L.
[1] "Cosmological models in differential geometry" University of Minnesota (1963) (mimeographed).

Marcus, M., and Mizel, V.
[1] Functional Composition of Sobolev Spaces, Bull. Am. Math. Soc. 78 (1972) 38-42.

Marsden, J.
[1] Hamiltonian one parameter groups, Arch. Rat. Mech. and Anal. 28 1968 (362-396).
[2] Publications du Departement de Mathematiques de Lyon, vol 4 (1972) 194-207.
[3] The Hopf bifurcation for nonlinear semigroups, Bull. Am. Math. Soc. 79 (1973) 537-541.
[4] Darboux's Theorem Fails for Weak Symptectic Forms, Proc. Am. Math. Soc. 32 (1972) 590-592.
[5] On Product Formulas for Nonlinear Semigroups Journ. Funct. An. 13 (1973) 51-72.
[6] A Formula for the Solution of the Navier-Stokes Equations, based on a Method of Chorin, Bull. Am. Math. Soc. 80 (1974).
[7] Some Fundamental Properties of the Solutions of the Euler and Navier-Stokes Equations (in preparation).
[8] A proof of the Calderon Extension Theorem, Can. Math. Bull. 16 (1973) 133-136.
[9] On Completeness of Homogeneous Pseudo Riemannian Manifolds, Indiana Univ. Math. Journ. 22 (1973) 1065-1066.
[10] On Global Solutions for Nonlinear Hamiltonian Evolution Equations, Comm. Math. Phys. 30 (1973) 79-81.

Marsden, J., and Abraham, R.
[1] Hamiltonian Mechanics on Lie Groups and Hydrodynamics, Proc. Pure Math XVI, Amer. Math. Soc. (1970) 237-243.

Marsden, J., Ebin, D., and Fischer, A.
[1] Diffeomorphism Groups, Hydrodynamics and Relativity, Proc. 13th biennial seminar of Can. Math. Congress, ed. J.R. Vanstone, Montreal (1972) 135-279.

Marsden, J., and Fischer, A.
[1] General Relativity as a Hamiltonian System, in Symposia Mathematica, Academic Press (to appear).

Marsden, J., and Weinstein, A.
[1] Reduction of Symplectic Manifolds with Symmetry, Reports on Math. Phys. (to appear).

Martin, M.H.
[1] The flow of a viscous fluid, I, Arch. Rat. Mech. Anal. 41 (1971) 226-286.

Maslov, V.P.
[1] "Théorie des Perturbations et Méthodes Asymptotiques," Dunod (1972).

Mather, J.
[1] Appendix of Smale [2].

McGrath, F.J.
[1] Nonstationary plane flow of viscous and ideal fluids, Arch. Rat. Mech. Anal. 27 (5) (1968), 329.

McLeod, J.B., and Serrin, J.
[1] The Existence of Similar Solutions for some Laminar Boundary Layer Problems, Arch. Rat. Mech. Anal. 31 (1969), 288-303.

McLeod, J.B., and Sattinger, D.H.
[1] Loss of Stability and bifurcation at a double eigenvalue, J. Funct. An. 14 (1973) 62-84.

Meyer, K.R.
[1] Generic Bifurcation of Periodic Points, Trans. Am. Math. Soc. 149 (1970) 95-107.

Milnor, J.
[1] "Morse Theory," Princeton University Press (1963).

Misner, C.
[1] Feynman Quantization of General Relativity, Rev. Mod. Phys. 29 (1957) 497-509.

Montgomery, D.
[1] On Continuity in Topological Groups, Bull. Am. Math. Soc. 42 (1936) 879.

Morrey, C. Jr., and Eells, J.
[1] A Variational Method in the Theory of Harmonic Integrals 63 (1956) 91-128.

Morrey, C.B., Jr.
[1] A Variational Method in the Theory of Harmonic Integrals II, Am. J. Math. 78 (1956) 137-170.
[2] "Multiple Integrals in the Calculus of Variations," Springer, 1966.

Moser, J.
[1] On the Volume Elements on a Manifold, Trans. of Amer. Math. Soc. 120 (1965), 286-294.

Naimark, J.
[1] Dokl Akad. Nauk SSR 129 (1959) No. 4
[2] Motions Close to Doubly Asymptotic Motion, Soviet Math. Dokl. 8 (1967) 228-231.

Nash, J.
[1] Le probleme de Cauchy pour les equations differentielles d'un fluide general, Bull. Soc. Math. France 90 (1962), 487-497.

Nelson, E.
[1] "Topics in Dynamics I, Flows," Princeton University Press (1969).
[2] "Dynamical Theories of Brownian Motion" Princeton University Press (1967).
[3] Feynman Integrals and the Schrödinzer Equation, J. Math. Phys 5 (1964) 332-343.

VonNeumann, J.
[1] Recent Theories of Turbulence, Collected Works, VI, Macmillan, N.Y. 1963, 437-472.
[2] "Mathematical Foundations of Quantum Mechanics," Princeton Univ. Press (1955).

Nirenberg, L.
[1] On Elliptic Partial Differential Equations, Annali delli Scuola Norm. Sup., Pisa 13 (1959) 115-162.

Omori, H.
[1] On the Group of Diffeomorphisms on a Compact Manifold, in Proc. Symp. Pure Math. XV, Amer. Math. Soc. (1970) 167-184.
[2] Local structures of groups of diffeomorphisms, J. Math. Soc. Japan 24 (1972) 60-88.
[3] On Smooth extension theorems, J. Math. Soc. Japan 24 (1972) 405-432.

N. O'Murchadha and J.W.York
[1] Existence and uniqueness of solutions of the Hamiltonian constraint of gereral relativity on compact manifolds. J. Math. Phys. 14(1973)1551-7.

Orszag, S.A.
 [1] Analytical theories of turbulence. J. Fluid Mech. 41 (1970) 363-386.

Palais, R.
 [1] "Seminar on the Atiyah-Singer Index Theorem," Princeton, 1965.
 [2] On the Homotopy Type of Certain Groups of Operators, Topology 3 (1965) 271-279.
 [3] Homotopy Theory of Infinite Dimensional Manifolds, Topology 5 (1966) 1-16.
 [4] "Foundations of Global Non-Linear Analysis," Benjamin, N.Y. 1968.
 [5] The Morse Lemma on Banach Spaces, Bull. Am. Math. Soc. 75 (1969) 968-971.
 [6] Extending Diffeomorphisms, Proc. Am. Math. Soc. 11 (1960) 274-277.
 [7] Morse Theory on Hilbert Manifolds, Topology 2 (1963) 299-340.

Payne, L.E., and Weinberger, H.F.
 [1] An Exact Stability Bound for Navier-Stokes Flow in a Sphere, in "Non-Linear Problems," ed. Langer, University of Wisconsin Press (1962).

Penot, J.P.
 [1] Variete differentiables d'applications et de chemins. C.R. Acad. Sc. Paris 264 (1967) 1066-1068.
 [2] Une methode pour construire des varietes d'applications au moyen d'un plongement, C.R. Acad. Sc. Paris 266 (1968) 625-627.
 [3] Geometrie des varietes fonctionnelles, These, Paris (1970).
 [4] Sur la theorem de Frobenius, Bull. Math. Soc. France 98 (1970) 47-80.
 [5] Topologie faible sur des varietes de Banach C.R. Acad. Sc. Paris 274 (1972) 405-408.

Petrovskii, I.
 [1] Uber das Cauchysche problem fur lineare und nichtlineare hyperbolische partielle Differentialgleichungen, Rec. Math. (math. Sbornik), N.S. 2, 44 (1937) 814-868.

Pirani, F.
 [1] "Lectures on General Relativity," Brandeis Summer Institute of Physics, Volume One (1964).

Poincaré, H.
 [1] "Les methodes nouvelles de la mechanique celeste", Gauthier-Villars, Paris (1892).

Ponomarenko, T.B.
 [1] Occurrence of space-periodic motions in hydro-
     dynamics, J. Appl. Math Mech. 32 (1968) 40-51, 234-245.

Quinn, B.
 [1] Solutions with Shocks: An Example of an L_1-Contractive
     Semigroup, Comm. Pure and Appl. Math. XXIV (1971)
     125-132.

Rabinowitz, P.H.
 [1] Existence and Nonuniqueness of Rectangular Solutions
     of the Benard Problem, Arch. Rat. Mech. Anal. 29
     (1967) 30-57.
 [2] Some Global Results for Non-linear Eigenvalue
     Problems, J. Funct. Anal. 7 (1971) 487-513.

Reynolds, O.
 [1] On the Dynamical Theory of Incompressible Viscous
     Fluids and the Determination of the Criterion, Phil.
     Trans. Roy. Soc. London A186 (1895) 123-164.

Riddell, R.C.
 [1] A Note on Palais' axioms for section functors, Proc.
     Am. Math. Soc. 25 (1970) 808-810.

Robbin, J.
 [1] On the Existence Theorem for Differential Equations,
     Proc. Amer. Math. Soc. 19 (1968) 1005-1006.
 [2] Stable Manifolds of Hyperbolic Fixed Points, Illinois
     Journ. Math. 15 (1971) 595-609.
 [3] A Structural Stability Theorem, Ann. of Math. 94
     (1971) 447-493.
 [4] Relative Equilibria in Mechanical Systems, Publ. Dept.
     Math. Univ. de Lyon, Vol, 4 (1972) 136-143.

Roels, J., and Weinstein, A.
 [1] Functions whose Poisson Brackets are Constants,
     J. Math. Phys. 12 (1971), 1482-1486.

Rosenblat, M., and Van Atta, C. (ed.)
 [1] "Statistical Models and Turbulence," Springer Lecture
     Notes in Physics #12 (1972).

Roseau, M.
 [1] Vibrations non linéares et théorie de la stabilité,
     Springer Tracts in Nat. Phil., No. 8 (1966).

Ruelle, D.
 [1] Statistical Mechanics on a compact set with a Z^ν
     action satisfying expansiveness and specification
     Publ. du dept. math. Univ. Lyon (1972), and
     Bull. Am. Math. Soc. 78 (1972) 988-991.

Ruelle, D. (Cont'd.)
 [2] Dissipative systems and differential Analysis, Boulder Lectures (1971)
 [3] Strange Attractors as a Mathematical Explanation of Turbulence, pp. 292-299 of Rosenblatt-Van Atta [1].
 [4] Bifurcations with Symmetries, Arch. Rat. Mech 51 (1973) 136-152.
 [5] "Statistical Mechanics - Rigorous Results" W.A. Benjamin (1969).
 [6] A measure associated with an Axiom A attractor (to appear).

Ruelle, D., and Takens, F.
 [1] On the Nature of Turbulence, Comm. Math. Phys. 20 (1971) 167-192, 23 (1971) 343-344.

Rund, H., and Lovelock, D.
 [1] Variational Principles in the general theory of relativity, Jber. Deutsch. Math - Verein. 74 (1972) 1-65.

Saber, J.
 [1] On Manifolds of Maps, Thesis, Brandeis University (1965).

Sachs, R., and Wu, H.
 [1] "Basic Relativity for students of Mathematics" (in preparation).

Sattinger, D.H.
 [1] Bifurcation of Periodic Solutions of the Navier-Stokes Equations, Arch. Rat. Mech. Anal. 41 (1971) 66-80.
 [2] On Global Solution of Nonlinear Hyperbolic Equations, Arch. Rat. Mech. Anal. 30 (1968) 148-172.
 [3] The Mathematical Problem of Hydrodynamic Stability, J. Math. and Mech. 19 (1971) 797-817.
 [4] "Topics in Stability and Bifurcation Theory" Springer lecture notes #309 (1973).
 [5] Transformation groups and bifurcation at multiple eigenvalues, Bull. Am. Math. Soc. 79 (1973) 709-711.

Schauder, J.
 [1] Das Anfangswertproblem eimer quasilinearen hyperbolischen Differentialgleichungen zweiter Ordnung in beliebiger Anzahl von unabhangigen Veranderlichen, Fund. Math. 24 (1935) 213-246.

Schnute, J., and Shinbrot, M.
 [1] The Cauchy Problem for the Navier-Stokes Equations (preprint).

Schwartz, J.T.
[1] "Nonlinear Functional Analysis" Gordon and Breach (1967).

Segal, I.
[1] Nonlinear Semigroups, Ann. of Math. 78 (1963) 339-364.
[2] Differential Operators in the Manifold of Solutions of a Nonlinear Differential Equation, Journ. Math. Pures et Appl., XLIV (1965) 71-113.
[3] "Mathematical Problems of Relativistic Physics" Am. Math. Soc. (1963).
[4] "A Mathematical Approach to Elementary particles and their Fields" Univ. of Chicago lecture notes (1955).

Sel'kov, E.
[1] Self-Oscillations in Glycose, Europ. Journ. Biochem. 4 (1968) 79-86.

Serrin, J.
[1] Mathematical Principles of Classical Fluid Dynamics, Encyclopedia of Physics, Vol. 8/1, Springer (1959).
[2] On the Stability of Viscous Fluid Motions, Arch. Rat. Mech. Anal. 3 (1959) 1-13.
[3] The Initial Value Problem for the Navier-Stokes Equations in "Non Linear Problems," ed. Langer, University of Wisconsin Press (1962).
[4] "Mathematical Aspects of Boundary Layer Theory," Lecture Notes, Univ. of Minnesota, 1962 (out of print).
[5] On the Interior Regularity of Weak Solutions of the Navier-Stokes Equations, Arch. Rat. Mech. Anal. 7 (1962) 187-195.
[6] On the Mathematical Basis for Prandtl's Boundary Layer Theory: An Example, Arch. Rat. Mech. Anal. 28 (1968) 217-225.

Shinbrot, M.
[1] Fractional Derivatives of Solutions of the Navier-Stokes Equations, Arch. Rat. Mech. Anal. 40 (1971) 139-154.

Simmons, G.
[1] "Introduction to Topology and Modern Analysis" McGraw-Hill (1963).

Simms, D.
[1] "Lie Groups and Quantum Mechanics," Springer Lecture Notes #52 (1968).
[2] Geometric quantization of Energy Levels in the Kepler Problem, Symposia Mathematica (to appear).

Simon, B.
 [1] Selected Topics in Functional Analysis, in "Mathematics of Contemporary Physics" R. Streater, ed. Acad. Press, N.Y. 1972.

Sinai, Ya G.
 [1] Measures invariantes des Y-systemes, Actes Congres Intern. Math. tome 2 (1970) 929-940.

Smale, S.
 [1] Diffeomorphisms of the Two Sphere, Proc. Amer. Math. Soc. 10 (1959) 621-626.
 [2] Differentiable Dynamical Systems, Bull. Am. Math. Soc. 73 (1967) 747-817.
 [3] Morse Theory and a Nonlinear Generalization of the Dirichlet Problem, Ann. of Math. 80 (1964) 382-396.
 [4] Topology and Mechanics (I,II), Inv. Math. 10 (1970) 305-331, 11 (1970), 45-64.

Sniatycki, J., and Tulczyjew, W.H.
 [1] Generating forms for Lagrangian Submanifolds, Indiana Math. Journ 22 (1972) 267-275.
 [2] Canonical Dynamics of Relativistic Charged Particles, Ann. Inst. H. Poincare 15 (1971) 177-187.

Sobolev, S.L.
 [1] "Applications of Functional Analysis in Mathematical Physics," Translations of Math. Monographs, Vol. 7, Am. Math. Soc., Providence, R.I. (1963).

Sobolevskii, P.E.
 [1] An Investigation of the Navier-Stokes Equations by Means of the Theory of Parabolic Equations in Banach Spaces, Sov. Math. Dokl. 5 (1964) 720-723.

Souriau, J.M.
 [1] "Structure des Systemes Dynamiques" Dunod, Paris (1970).
 [2] "Géometric et relativité," Hermann, Paris (1964).
 [3] Sur la varieté de Kepler, Symposia Mathematica (to appear).

Sternberg, S.
 [1] Lectures on Differential Geometry, Prentice Hall, Englewood Cliffs, N.J. 1963.

Swann, H.S.G.
 [1] The Convergence with Vanishing Viscosity of Nonstationary Navier-Stokes Flow to Ideal Flow in R^3, Trans. A.M.S. 157 (1971) 373-397.

Synge, J.L.
 [1] "Relativity, The General Theory," North-Holland (1960).

Synge, J.L., and Griffith, B.A.
[1] "Principles of Mechanics," McGraw-Hill (1959)

Taub, A.H.
[1] On Hamilton's Principle for Perfect Compressible Fluids, Proc. Symp. Appl. Math. I, Amer. Math. Soc. (1949) 148-157.

Taylor, E.F., and Wheeler, A.
[1] "Spacetime Physics," W.H. Freeman, San Francisco (1966).

Taylor, G.I.
[1] Statistical Theory of Turbulence, Parts 1-4, Proc. Roy. Soc. 151 (1935) 421.

Temam, R.
[1] Une methode d'approximation de la solution des equation de Navier-Stokes Equations, Bull. Soc. Math. France (1970).
[2] Quelques methodes de decomposition en analyse numerique, Actes Congres Intern. Math. (1970) tome 3, 311-319.

Ton, B.A.
[1] Strongly Nonlinear Parabolic Equations, J. Funct. Anal. 7 (1971) 147-155.

Tromba, A.J.
[1] The Morse Lemma on Arbitrary Banach Spaces, Bull. Am. Math. Soc. 79 (1973) 85-86.

Trotter, H.F.
[1] On the Product of Semi-group of Operators, Proc. Amer. Math. Soc. 10 (1959) 545-551.

Turing, A.M.
[1] The Chemical Basis of Morphogenesis, Phil. Trans. Roy. Soc. (B) 237 (1952) 37-72.

Varadarajan, V.S.
[1] "Geometry of Quantum Theory (I,II)," Van Nostrand (1968).

Velte, W.
[1] Uber ein Stabilitatskiterium der Hydrodynamik, Arch. Rat. Mech. Anal. 9 (1962) 9-20.
[2] Stabilitats verhalten und Verzweigung stationaret Losungen der Navier Stokesschen Gleichungen, Arch. Rat. Mech. An. 16 (1964) 97-125.

Velte, W. (Cont'd.)
[3] Stabilitat und verzweigung stationaret Losungen der Navier-Stokesschen Gleichungen beim Taylorproblem, Arch. Rat. Mech. Anal. 22 (1966) 1-14.

Warner, F.
[1] "Foundations of Differentiable Manifolds and Lie Groups," Scott, Foresman (1971).

Weinberg, S.
[1] "Gravitation and Cosmology" Wiley, N.Y. (1972).

Weinstein, A.
[1] Symplectic Manifolds and their Lagrangian Submanifolds, Advances in Math., 6 (1971) 329-346. See also Bull. Am. Math. Soc. 75 (1969) 1040-1041.
[2] Quasi-classical mechanics on spheres, Symposia Mathematica (to appear).

Weinstein, A., and Marsden, J.
[1] A Comparison Theorem for Hamiltonian Vector Fields, Proc. Am. Math. Soc. 26 (1970) 629-631.

Weyl, H.
[1] The method of Orthogonal Projection in Potential Theory, Duke Math. J. 7 (1940) 411-444.
[2] "Space-Time-Matter" 4th ed., Dover (1922).

Wheeler, J.A.
[1] Geometrodynamics and the Issue of the Final State, in "Relativity, Groups and Topology," ed. DeWitt and DeWitt, Gordon and Breach, New York, 1964.
[2] Problems on the frontiers between general relativity and differential geometry, Rev. Mod. Phys 34 (1962) 873-892.
[3] "Geometrodynamics," Academic Press (1962).

Whittaker, E.T.
[1] Analytical Dynamics, 4th ed., Cambridge Univ. Press (1937).

Whitney, H.
[1] On the extension of differentiable functions, Bull. Am. Math. Soc. 50 (1944) 76-81.

Williams, R.F.
[1] One Dimensional Non-Wandering Sets, Topology 6 (1967) 473-487.

Wolf, J.A.
[1] "Spaces of Constant Curvature", 2nd edition, J.A. Wolf, Berkeley, Calif. (1972).
[2] Isotropic manifolds of indefinite metric, Comment. Math. Helv. 39 (1964) 21-64.

Wolibner, W.
[1] Un théorème sur l'existence du mouvement plan d'un fluide parfait homogène, incompressible, pendant un temps infiniment longue, Math. Z, 37 (1933) 698-726.

Yano, K.
[1] "Integral Formulas in Riemannian Geometry," Marcel Dekker, N.Y. (1971).

Yosida, K.
[1] "Functional Analysis," Springer, 1965.

Zarantonello, E.H. (ed.)
[1] "Contributions to Nonlinear Functional Analysis" Acad. Press, N.Y. (1971).